Ernst Schering Research Foundation Workshop 24
Molecular Basis of Sex Hormone Receptor Function

Springer-Verlag Berlin Heidelberg GmbH

Ernst Schering Research Foundation
Workshop 24

Molecular Basis of Sex Hormone Receptor Function

New Targets for Intervention

H. Gronemeyer, U. Fuhrmann, K. Parczyk
Editors

With 49 Figures

Springer

ISSN 0947-6075
ISBN 978-3-662-03691-4

CIP data applied for

Die Deutsche Bibliothek – CIP-Einheitsaufnahme
Schering-Forschungsgesellschaft <Berlin>: Ernst Schering Research Foundation Workshop.

ISSN 0947-6075
24. Molecular basis of sex hormone receptor function. - 1998
Molecular basis of sex hormone receptor function: new targets for intervention / H. Gronemeyer ... ed.

(Ernst Schering Research Foundation Workshop; 24)
ISBN 978-3-662-03691-4 ISBN 978-3-662-03689-1 (eBook)
DOI 10.1007/978-3-662-03689-1

Originally published by Springer-Verlag Berlin Heidelberg New York in 1998
Softcover reprint of the hardcover 1st edition 1998

Typesetting: Data conversion by Springer-Verlag

SPIN: 10657859 13/3135–5 4 3 2 1 0 – Printed on acid-free paper

Preface

The structures of steroid hormones are too simple to embody much information. The cellular receptors for the hormones must therefore serve as molecular adapters to mount a signal of considerable complexity.
[Bishop JM (1986) Oncogenes as hormone receptors. Nature 321: 112-113]

How steroid hormone receptors – and all of the other members of what is now recognized as the superfamily of nuclear receptors – can do such a remarkable job has fascinated scientists for decades. Indeed, in 1985 Keith R. Yamamoto, in an article in the *Annual Review of Genetics* (19:209-252) compared nuclear receptors with transfer RNAs, since both adapters interpret a code: Transfer RNAs confer upon amino acids the ability to "read" the genetic code, while nuclear receptors interpret a different sort of code specified by the molecular structure of the cognate ligand to initiate complex gene programs, regulating phenomena as diverse as pregnancy, bone density, energy metabolism, cell growth, and apoptosis. A series of discoveries over the past few years have now dramatically widened our understanding of the molecular mechanisms that govern nuclear receptor action. The elucidation of the three-dimensional structures of several nuclear receptor ligand-binding domains in the presence of agonists or antagonists has provided an allosteric concept of ligand action, and the discovery of a plethora of ligand-dependent protein interactions has linked this transconformation with the ability of nuclear receptors to act as transcriptional activators and repressors and the enzymatic modification of chromatin and factors of the basal transcriptional machinery.

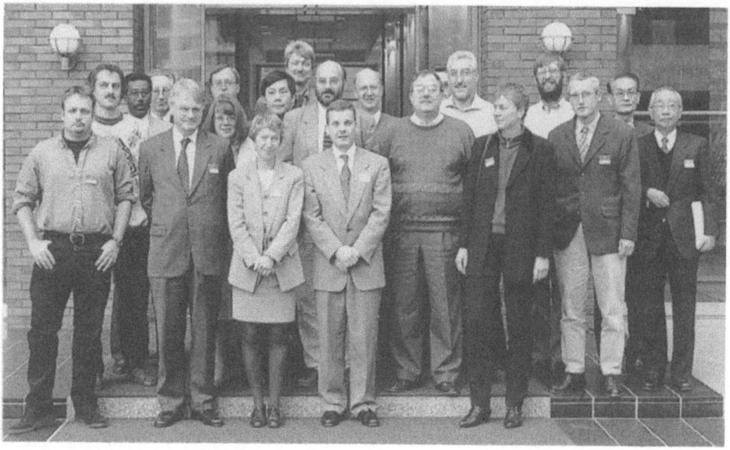

The participants of the workshop

The impact of recent developments in the nuclear receptor field reaches far beyond the innocent towers of basic science: The exciting perspectives of novel types of nuclear receptor-based drugs for a variety of diseases, ranging from osteoporosis, diabetes, rheumatoid arthitis, and obesity to various types of cancer, have electrified the managements of all major pharmaceutical companies. And there is more to come: Novel ligands are being identified for old "orphans," expanding the regulatory power of the nuclear receptor system, while the discovery of novel receptors, such as the estrogen receptor β, suggest that it may be possible to separate estrogen action into distinct, and perhaps cell-specific, events by ligand design.

The aim of the Ernst Schering Research Foundation Workshop 24 was to bring together leading nuclear receptor experts to present their latest findings and to discuss the current view of the "molecular basis of sex hormone receptor function". This workshop has been an exciting endeavor for the organizers and we hope to confer the enthusiasm of all participants for the study of nuclear receptor action on the reader of this monograph.

H. Gronemeyer, U. Fuhrmann, and K. Parczyk

Table of Contents

List of Editors and Contributors

Editors

H. Gronemeyer
Institute of Genetics and Molecular and Cell Biology, Parc d'Innovation, 1, rue Laurent Fries, B.P. 163, 67404 Illkirch Cedex, France

U. Fuhrmann
Fertility Control and Hormone Therapy, Schering AG, Müllerstr. 178, 13342 Berlin, Germany

K. Parczyk
Experimental Oncology, Schering AG, Müllerstr. 178, 13342 Berlin, Germany

Contributors

C. Anderson
Metabolic Research Unit, University of California, 1141 HSW, San Francisco, CA 94143-0540, USA

A. C. B. Cato
Center for Nuclear Research KFA Karlsruhe, Institute of Genetics, P.O. Box 3640, 76021 Karlsruhe, Germany

C. M. W. Chan
Academic Department of Biochemistry, The Royal Marsden NHS Trust, Fulham Road, London SW3 6JJ, United Kingdom

C. Chang
George Whipple Laboratory for Cancer Research, Departments of Pathology and Laboratory Medicine, University of Rochester Medical Center, 601 Elmwood Avenue, Rochester, NY 14642, USA

O. Donzé
Department of Cell Biology, University of Geneva, Sciences III, 30, Quai Ernest-Ansermet, 1211 Geneva 4, Switzerland

M. Dowsett
Academic Department of Biochemistry, Royal Marsden Hospital, Fulham Road, London SW3 6JJ, United Kingdom

E. Enmark
Center for Biotechnology and Department of Medical Nutrition, Karolinska Institute, Huddinge University Hospital, NOVUM, 14157 Huddinge, Sweden

N. Fujimoto
George Whipple Laboratory for Cancer Research, Departments of Pathology, Urology, and Biochemistry, University of Rochester Medical Center, 601 Elmwood Avenue, Rochester, NY 14642, USA

P. Giangrande
Department of Pharmacology, Dulx University Medical Center, Box 3813, Durham, NC 27710, USA

H. Gronemeyer
Institute of Genetics and Molecular and Cell Biology, Parc d'Innovation, 1, rue Laurent Fries, B.P. 163, 67404 Illkirch Cedex, France

J.-Å. Gustafsson
Department of Medical Nutrition, Karolinska Institute, Huddinge University Hospital, NOVUM, 14157 Huddinge, Sweden

D. Heery
Molecular Endocrinology Laboratory, Imperial Cancer Research Fund, 44 Lincoln's Inn Fields, London WC2A 3PX, United Kingdom

M. J. S. Heine
Institute of Genetics and Molecular and Cell Biology, Parc d'Innovation,
1, rue Laurent Fries, B.P. 163, 67404 Illkirch Cedex, France

P. Hsiao
George Whipple Laboratory for Cancer Research, Departments of Pathology,
Urology, and Biochemsitry, University of Rochester Medical Center,
601 Elmwood Avenue, Rochester, NY 14642, USA

S.-B. Hwang
George Whipple Laboratory for Cancer Research, Departments of Pathology,
Urology, and Biochemistry, University of Rochester Medical Center,
601 Elmwood Avenue, Rochester, NY 14642, USA

S. Inui
George Whipple Laboratory for Cancer Research, Departments of Pathology,
Urology, and Biochemistry, University of Rochester Medical Center,
601 Elmwood Avenue, Rochester, NY 14642, USA

E. Kalkhoven
Molecular Endocrinology Laboratory, Imperial Cancer Research Fund,
44 Lincoln's Inn Fields, London WC2A 3PX, United Kingdom

H. Kang
George Whipple Laboratory for Cancer Research, Departments of Pathology,
Urology, and Biochemistry, University of Rochester Medical Center,
601 Elmwood Avenue, Rochester, NY 14642, USA

M.-C. Keightley
Department of Cell Biology, Baylor College of Medicine, 1 Baylor Plaza,
Houston, TX 77030, USA

M.-R. Keneally
Metabolic Research Unit, University of California, 1141 HSW,
San Francisco, CA 94143-0540, USA

B. P. Klaholz
Laboratory of Structural Biology, Institute of Genetics and Molecular
and Cell Biology, CNRS/INSERM/ULP, Parc d'Innovation,
1, rue Laurent Fries, B.P. 163, 67404 Illkirch Cedex, France

P. J. Kushner
Metabolic Research Unit, University of California, 1141 HSW,
San Francisco, CA 94143-0540, USA

D. P. McDonnell
Department of Pharmacology, Dulx University Medical Center, Box 3813,
Durham , NC 27710, USA

H. Miyamoto
George Whipple Laboratory for Cancer Research, Departments of Pathology,
Urology, and Biochemistry, University of Rochester Medical Center,
601 Elmwood Avenue, Rochester, NY 14642, USA

D. Moras
Laboratory of Strucural Biology, Institute of Genetics and Molecular
and Cell Biology, CNRS/INSERM/ULP, Parc d'Innovation,
1, rue Laurent Fries, B.P. 163, 67404 Illkirch Cedex, France

L. V. Nazareth
Department of Cell Biology, Baylor College of Medicine, 1 Baylor Plaza,
Houston, TX 77030, USA

K. Nishimura
George Whipple Laboratory for Cancer Research, Departments of Pathology,
Urology, and Biochemistry, University of Rochester Medical Center,
601 Elmwood Avenue, Rochester, NY 14642, USA

K. Paech
Departments of Pharmaceutical Chemistry and Molecular Pharmacology,
University of California, San Francisco, CA 94143-0540, USA

M. G. Parker
Molecular Endocrinology Laboratory, Imperial Cancer Research Fund,
44 Lincoln's Inn Fields, London WC2A 3PX, United Kingdom

D. Picard
Departement of Cell Biology, University of Geneva, Sciences III,
30, Quai Ernest-Ansermet, 1211 Geneva 4, Switzerland

G. Pollio
Institute of Pharmacological Siences, University of Milan, Via Balzaretti 9,
20133 Milano, Italy

T. S. Scanlan
Departments of Pharmaceutical Chemistry and Molecular Pharmacology,
University of California, San Francisco, CA 94143-0540, USA

J. Shinsako
Metabolic Research Unit, University of California, 1141 HSW,
San Francisco, CA 94143-0540, USA

H. Ting
George Whipple Laboratory for Cancer Research, Departments of Pathology,
Urology, and Biochemistry, University of Rochester Medical Center,
601 Elmwood Avenue, Rochester, NY 14642, USA

J. Trapman
Department of Pathology, Erasmus University, P.O. Box 1738,
Rotterdam 3000 DR, The Netherlands

H. Uemura
George Whipple Laboratory for Cancer Research, Departments of Pathology,
Urology, and Biochemistry, University of Rochester Medical Center,
601 Elmwood Avenue, Rochester, NY 14642, USA

R. Uht
Metabolic Research Unit, University of California, 1141 HSW,
San Francisco, CA 94143-0540, USA

J. Valentine
Molecular Endocrinology Laboratory, Imperial Cancer Research Fund,
44 Lincoln's Inn Fields, London WC2A 3PX, United Kingdom

C. Wang
George Whipple Laboratory for Cancer Research, Departments of Pathology,
Urology, and Biochemistry, University of Rochester Medical Center,
601 Elmwood Avenue, Rochester, NY 14642, USA

P. Webb
Metabolic Research Unit, University of California, 1141 HSW,
San Francisco, CA 94143-0540, USA

N. L. Weigel
Department of Cell Biology, Baylor College of Medicine, 1 Baylor Plaza,
Houston, TX 77030, USA

S. Yeh
George Whipple Laboratory for Cancer Research, Departments of Pathology,
Urology, and Biochemistry, University of Rochester Medical Center,
601 Elmwood Avenue, Rochester, NY 14642, USA

Y. Zhang
Department of Cell Biology, Baylor College of Medicine, 1 Baylor Plaza,
Houston, TX 77030, USA

1 Nuclear Receptors as Targets for Drug Design: New Options and Old Challenges

M.J.S. Heine and H. Gronemeyer

1.1 Introduction

1.1.1 The Present Status: Nuclear Receptor Ligands in Therapy

Steroid receptors, a subgroup of the superfamily of nuclear receptors (Gronemeyer and Laudet 1995), have been the target for pharmacological drug design since their recognition as "master genes" with pleiotropic action on various physiological processes (for a review see De Groot 1995). All nuclear receptors (NRs) are transcription factors regulating specific gene programs in response to binding of their cognate ligands through alteration of: (a) the activity of the basal transcription machinery, and (b) chromatin structure at target genes (Fig. 1). In addition to this "direct" action, NRs can also mutually interfere, positively or negatively, with a number of other signaling processes (a poorly understood phenomenon, often referred to as "signal transduction crosstalk"; Fig. 1). Finally, as is increasingly recognized, NRs themselves can be targets for other signaling pathways and can be modified post-transcriptionally at specific sites, thereby modulating receptor activity.

The enormous impact of research into NRs and NR-mediated drug action on human life and society is obvious from the role of contraceptive and abortifacient steroids, anti-inflammatory glucocorticoids, thyroid hormones in growth defect treatments, and the cancer-preventive (for recent advances in chemoprevention of cancer see Hong and Sporn 1997) and cancer-therapeutic action of NR ligands. This (non-exhaustive) list of current medical therapies exploiting or interfering with NR activities also includes endocrine cancer therapies which inhibit the growth-promoting activity of the estrogen and androgen receptors in breast and prostate cancer, respectively, or induce the differentiative, anti-proliferative and apoptotic activity of retinoid receptors (for an extensive discussion on retinoids see Sporn et al. 1994), as in acute promyelocytic leukemia (Warrell et al. 1991). In addition, hormone

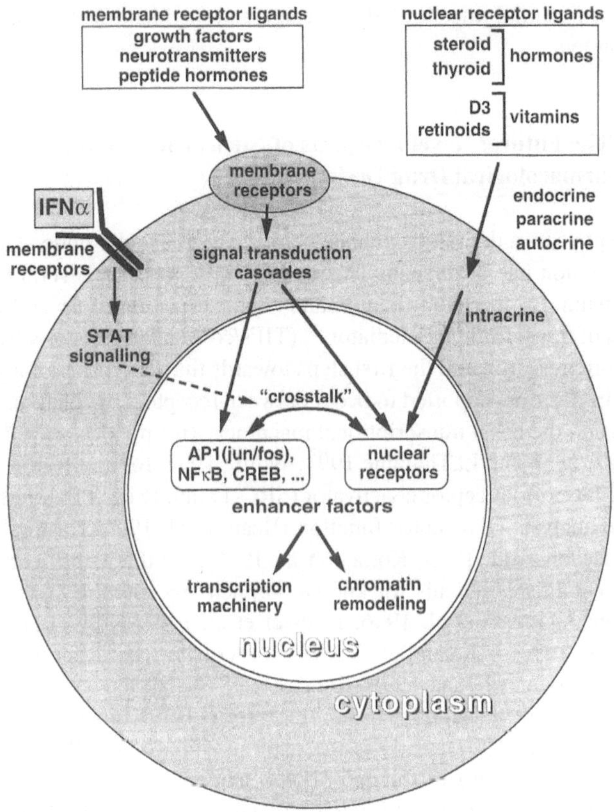

Fig. 1. Three major signal transduction pathways involving membrane or nuclear receptors. Note the "crosstalk" which gives rise to mutual interference between the various signaling cascades. *IFNα*, interferon α; *CREB*, cyclic adenosine monophosphate response element binding protein; *NFκB*, nuclear factor κB

replacement therapy in postmenopausal women is becoming increasingly important in aging societies.

From the above it is clear that some NRs have already proven to be prime targets for pharmaceutical drug development. Therefore, advances in the rapidly developing field of NR research must be evaluated for their potential impact on drug design and the identification of novel drug targets.

1.1.2 The Future: Novel Prospects of Nuclear Receptors for Pharmacological Drug Design

In recent years a dramatic enhancement of our molecular understanding of NR action has been seen. This is due to progress on several levels, comprising: (a) the identification and characterization of several novel classes of transcriptional mediators[1] (TIFs/co-regulators[2]; co-activators and co-repressors), and the first steps towards the decryption of a plethora of interactions reported to occur between receptors, mediators, chromatin, and the basal transcriptional machinery (for reviews see Glass et al. 1997; Lyon and LaThangue 1997; Wolffe 1997; for an illustration of the TIF2/steroid receptor coactivator (SRC-1) family see Fig. 2); (b) the genetic analysis of receptor function (Beato et al. 1995; Kastner et al. 1995; Lydon et al. 1995; Korach et al. 1996); (c) the identification of novel (candidate) ligands for known "orphan" receptors (Devchand et al. 1996; Janowski et al. 1996; Kliewer et al. 1997; Krey et al. 1997; Lala et al. 1997; Lehmann et al. 1997; for a review on earlier work, see Mangelsdorf and Evans 1995; Enmark and Gustafsson 1996); (d) the identification of a second estrogen receptor (ERβ; Kuiper et al. 1996; Tremblay et al. 1997); and (e) the determination of the three-dimensional (3D) structures of the apo-, holo-, and antagonist-bound ligand-binding domains (LBDs) of several NRs (Bourguet et al. 1995; Renaud et al. 1995; Wagner et al. 1995; Brzozowski et al. 1997), together with the prediction of a common fold of all NR LBDs (Wurtz et al. 1996).

The importance of these achievements for the development of novel pharmaceutical drugs is obvious: The availability of 3D structures facili-

---

[1] The existence of transcription intermediary factors (TIFs/co-regulators) for NRs was originally predicted from "squelching" experiments (Bocquel et al. 1989; Meyer et al. 1989)

[2] Bona fide co-activators are SRC-1, Oñate et al. 1995; TIF2, Voegel et al. 1996, 1998; AIB1, Anzick et al. 1997; RAC3, Chen et al. 1997; ACTR, Li et al. 1997; TRAM-1, Takeshita et al. 1997; p/CIP, Torchia et al. 1997; bona fide co-repressors are N-CoR, Hörlein et al. 1995 and SMRT, Chen and Evans 1995; Chen et al. 1996a.

tates the design of new and more (isotype-)specific ligands, provides ideas about the structural basis of agonistic and antagonistic ligands, allows conclusions to be drawn about the details of ligand–receptor interactions and to verify them by receptor mutation and ligand modification, and provides a scaffold for the modeling of non-crystallized receptor LBDs. Transcriptional mediators and the interfaces established in the signal transduction cascades may be novel drug targets. Receptor isotype-specific ligands for the six retinoid (RAR and RXR α, β, and γ) and two estrogen (ERα and β) receptors have the potential to elicit selective, perhaps tissue-specific effects. This could enablethe direction of receptor activity to the affected tissue in an attempt to reduce side effects, as may be the case, for example, for ERβ and osteoporosis. Another potential way to reduce side effects is to generate "dissociated" ligands (Chen et al. 1995; Vayssière et al. 1997; and references therein). Ligand combinations for heterodimeric receptors may reduce effective doses (Chen et al. 1996b) and stimulation of particular pathways through the RXR partner may become an option (Mukherjee et al. 1997), once pathway-specific RXR ligands have been found. Novel ligands may reveal novel options for the treatment of diseases which were not recognized as being linked to the action of NRs. Furthermore, the pro- and anti-proliferative potential of NRs in certain tissues, as well as the potential role of co-regulator expression/amplification (Anzick et al. 1997), deserve significant attention. Indeed, advances in the chemoprevention of cancer through NRs have been reviewed recently (Hong and Sporn 1997).

In the following, we shall first review recent advances in the molecular understanding of NR function (progressing stepwise from LBD structure via co-regulator interaction towards a general model; see Sect. 1.2), then discuss several unresolved problems (Sects. 1.3 and 1.4), and finally reflect on novel options for drug development (Sect. 1.5).

1.2 Mechanisms of Action of Nuclear Receptors

1.2.1 The Mousetrap Model of LBD Structure

The first crystal structures of a non-liganded (apo-RXRα; Bourguet et al. 1995) and of an all-*trans* retinoic acid-bound (holo-RARγ; Renaud et al. 1995) retinoid receptor LBD demonstrated that the two LBDs have a common fold. Moreover, a structure-based alignment of all known NR LBDs provided a strong argument for a common fold for the LBDs of all members of the NR superfamily (Wurtz et al. 1996). This hypothesis was fully supported by the crystal structures of the holo-LBDs of the thyroid hormone (Wagner et al. 1995), estrogen (Brzozowski et al. 1997) and peroxisome proliferator-activated receptor γ (PPARγ) (Milburn et al. 1997) receptors. Consequently, we argued that the differences observed between apo-RXRα and holo-RARγ LBDs reflected the action of the ligand (Renaud et al. 1995). The most striking change was the transconformation of the 12th helix (H12), which encompasses the core of the ligand-inducible activation function-2 (AF-2) (for a review on NR AFs, see Gronemeyer and Laudet 1995; Chambon 1996). As is discussed in more detail by Renaud et al. (1995) and in Chap. 8, this volume, the apo- to holo- transconformation of the LBD resembles a mousetrap in which the "activation" helix H12 flips into an alternate position secured by the establishment of a novel salt bridge to a residue of helix H4. It is important to point out, however, that this is not the only conformational change, since further important conformational alterations occur, including the bending of helix H3, flipping of the Ω-loop, and a general compaction of the LBD. The consequence of this LBD transconformation is that the molecule displays a completely altered repertoire of potential surfaces for the interaction with other transcription factors.

1.2.2 What Springs the Trap?

Given the significant conformational change induced by ligand binding to the apo-LBD, it is an immediate question how the ligand can achieve this alteration[3]. When mutating the residues in the ligand-binding pocket (LBP) of RXRα in an attempt to reveal their role in ligand binding, we obtained an important hint. Mutating Phe318 of the mouse RXRα (corresponding to Phe313 of the human homologue) to alanine generated a receptor that was fully active in the absence of ligand. Using various types of assays we showed that mRXRαF313A exhibits structural and functional features that made it indistinguishable from the agonist-bound (holo-) wild-type RXRα (Vivat et al. 1997). When this mutation was modeled, we realized the presence of a network of van-der-Waals interactions in the apo-RXRα between LBP residues F313, I310 (H5), V349 (H7), I324 (β-turn) and the H11 residues F438 and L441. The mutation mRXRαF313A would disrupt this network which anchors H11 to the core of the LBD[3]. The release of H11 facilitates the disruption of the contacts between loop 11–12 and H12 with H3, such that H11 and H12 can rearrange towards a novel equilibrium which is dominated by the possibility to generate an important salt-bridge between H4 and H12 (Renaud et al. 1995; Vivat et al. 1997). In other words, the equilibrium between the apo- and holo-forms of RXRα is shifted towards the holo-form. Importantly, modeling also revealed that the presence of the ligand in the LBP is not compatible with the existence of the above-described van-der-Waals network, and we proposed that ligand binding causes a similar destabilization of H11 positioning resulting, as for the mRXRαF313 A mutation, in a shift of the equilibrium from the apo- to the holo-receptor (Vivat et al. 1997).

Mouse RXRαF318 is not conserved in other NRs and neither mutation of homologous residues nor mutations disrupting clusters of interactions similar to those in the apo-RXR LBP generate constitutively active RAR, PR, or ER mutants. Therefore, it is likely that other and/or additional interactions stabilize the apo- (and holo-) forms of these receptors. Interestingly, the generation of constitutively active retinoid X (Vivat et al. 1997) and estrogen (Weis et al. 1996; White et al. 1997) receptors by single amino acid residue mutations within and outside of their LBPs, respectively, suggests that post-translational modifications of NR LBD residues (e.g., by phosphorylation) could also shift the equilibrium between the apo and holo NR LBD structures in the absence of ligand, or alter this equilibrium in the presence of agonistic or antagonistic ligands. In this way, signals from other pathways could be

[3] The core of the LBD is the highly conserved "top" part comprising helices H3 to H10 (compare Wurtz et al. 1996)

transduced through NRs in the absence of cognate ligands or alter the activity of ligand-bound NRs.

1.2.3 Molecular Consequences of Ligand Binding

The structure of apo- and holo-NR LBDs per se would have invoked much less excitement without parallel studies asking what specific abilities a holo-LBD had acquired upon its transconformation. Evidence that NRs have indeed acquired a novel ability upon ligand binding was reported several years ago when we proposed a model according to which holo-NRs would transmit their activity to the basal transcription machinery via TIFs/co-regulators). This hypothesis was based on so-called squelching experiments in which it was observed that overexpression of one receptor ("autosquelching"; Bocquel et al. 1989) or of a different receptor ("heterosquelching"; Meyer et al. 1989) would inhibit agonist-induced transactivation in a squelching receptor dose- and ligand-dependent manner. These data were interpreted as the result of sequestration of TIFs from the transcription initiation complex by either excess of the same, or addition of another ligand-activated receptor (Bocquel et al. 1989; Meyer et al. 1989). This concept of TIFs/co-regulators which are shared among different NRs despite their considerable sequence divergence has been fully confirmed (see Sect. 1.2.4).

1.2.4 The Co-regulator Model for Transcriptional Activation and Silencing

1.2.4.1 Co-activators

Indeed, due to the development of the "two hybrid" and direct cDNA expression bank screening technologies a number of putative transcriptional mediators/TIFs, also called co-regulators, have been identified (for reviews, see Horwitz et al. 1996; Glass et al. 1997). Two distinct classes of TIFs exist: co-activators, which mediate the activity of the ligand-dependent NR AF-2 function, and co-repressors, which mediate the silencing function of (some) NRs.

Several groups have shown that some of these putative co-activators indeed exhibit characteristics which are expected from a bona fide TIF for AF-2: (a) It interacts in vivo with NRs in an agonist-dependent manner, (b) it binds directly to NR LBDs in an agonist- and AF-2-integrity-dependent manner in vitro, (c) it harbors an autonomous transcriptional activation function, (d) it relieves nuclear receptor autosquelching, and (e) it enhances the activity of some nuclear receptor AF-2s when overexpressed in mammalian cells. All of these properties are apparently displayed by what is now recognized as the TIF2/SRC-1 family of NR co-activators (schematically aligned in Fig. 2) which comprises TIF2/GRIP-1 (Voegel et al. 1996, 1998; Hong et al. 1996, 1997), SRC-1 (Spencer et al. 1997; Kamei et al. 1996) and p/CIP/RAC-3/AIB1/ACTR/TRAM-1 (Torchia et al. 1997; Li et al. 1997; Anzick et al. 1997; Chen et al. 1997; Takeshita et al. 1997). Also, CBP/p300, originally identified as co-activator of CREB, interacts directly, albeit rather weakly, with holo-NR LBDs (Kamei et al. 1996; see also Sect. 2.4.2 below). While all these bona fide co-activators display limited NR selectivity (see Sect. 1.3.5), specific co-activators have been identified for the androgen receptor (see Chap. 2, this volume). In addition to those mentioned above, several other putative NR co-activators have been reported, including TIF1α (Le Douarin et al. 1995), RIP140 (Cavaillès et al. 1995), RAP46 (Zeiner and Gehring 1995), SUG1/RIP1 (Lee et al. 1995; vom Baur et al. 1996), GRIP170 (Eggert et al. 1995), L7/SPA (Jackson et al. 1997). Whether these factors mediate their action at the transcriptional or other levels (e.g., SUG1 could be involved in co-transcriptional receptor degradation) remains to be established.

1.2.4.2 The TIF2/SRC-1 Family of Transcriptional Co-activators

The structure–function analyses by several groups have revealed the existence of a family of (at least) three structurally and functionally related co-activators, referred to here as the TIF2/SRC-1 family (Fig. 2). These proteins possess conserved NR interaction domains (NIDs) harboring three NR boxes (Le Douarin et al. 1996) which function in a redundant fashion (Voegel et al. 1998; Kalkhoven et al. 1998; for comparison see Fig. 3). NR boxes have also been observed in putative NR co-activators which do not belong to the TIF2/SRC-1 family (Le Douarin et al. 1996; Heery et al. 1997) and may represent archetypal interaction motifs, not necessarily confined to NR–co-activator interaction. SRC-1 is apparently distinct from the other family members in that it has a functional fourth NR box at the C-terminus of isoform SRC-1a

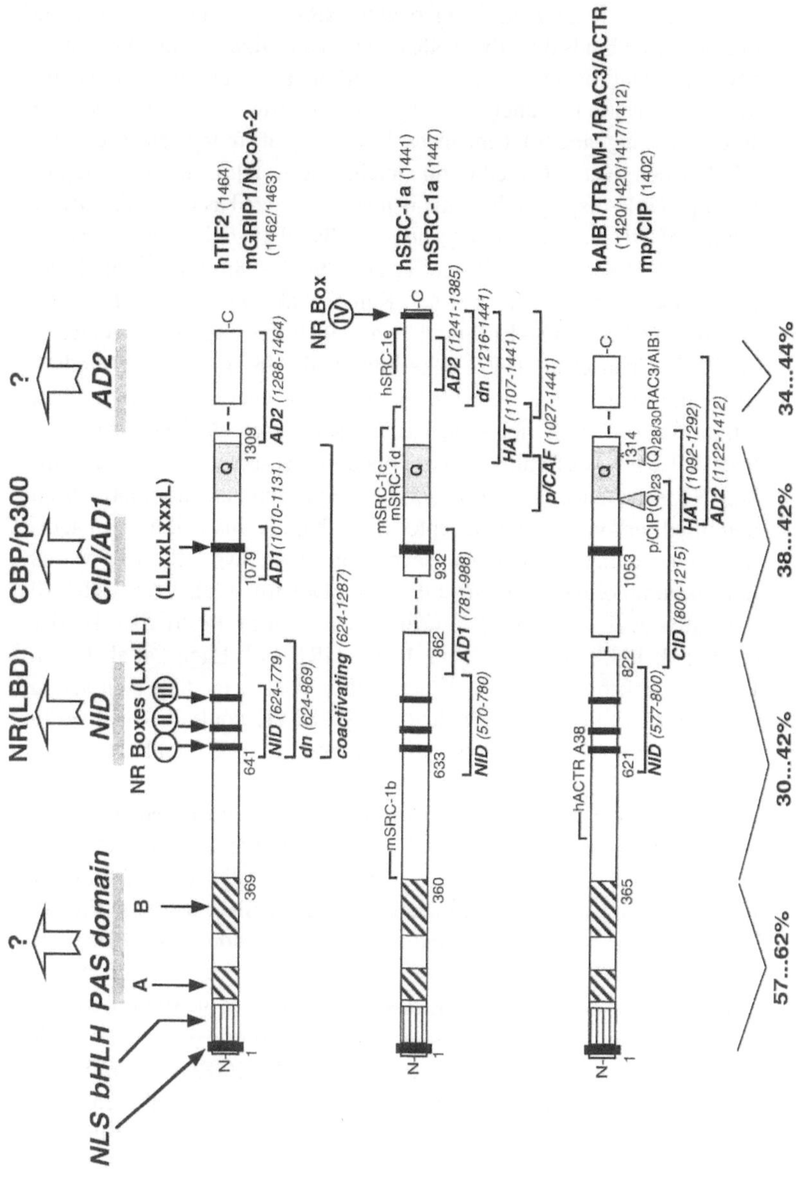

Fig. 2. Legend see p. 11

[see Chap. 3, this volume, and Kalkhoven et al. (1998) for the functional difference between SRC-1 isoforms]. The link between LBD H12 trans-conformation and TIF2 interaction has recently been provided by the crystallization of a 13-residue peptide of the TIF2 NR box 2 with the LBD of TRβ (B. Darimont, K.R. Yamamoto, R.L. Wagner, and R.J. Fletterick, personal communication). In solution this peptide has no detectable secondary structure, but in the complex with TRβ it forms an amphipathic α-helix. The hydrophobic surface of this helix, formed by the conserved hydrophobic residues of the NR box motif, interacts with

---

◀ **Fig. 2.** The transcriptional intermediary factor/steroid receptor coactivator (TIF2/SRC-1) family of nuclear receptor co-activators. Schematic alignment, drawn to scale, of the peptide sequences of the three known human orthologues TIF2, SRC-1a, and AIB1/RAC3/ACTR/TRAM-1. The amino acid lengths and the names of the respective murine homologues are indicated at the *right*. Isoforms (splicing variants) are represented by *square brackets* above the sequences, conserved motifs by *vertical bars* or *hatched boxes*, and alignment gaps by *dotted lines*. Polyglutamine insertions of different lengths are indicated for the bottom sequence; *Q*, glutamine-rich region. Note that the few significant peptide sequence differences between mp/CIP and hAIB1, including the distinct C-terminus, can be entirely attributed to reading frame shifts; thus mp/CIP probably refers to the (mutated) mouse homologue of the AIB1/RAC3/ACTR/TRAM-1 gene. Likewise, the mNCoA-2 sequence can be considered an allele of the mGRIP1 gene. Denoted at the *top* of the figure are the general features based on sequence homology (*NLS*, nuclear localization signal; *bHLH*, basichelix-loop-helix motif; *PAS*, Per-Arnt-Sim homology region) or functional evidence (*NID*, *CID*, nuclear receptor and CBP/p300 interaction domains, respectively; *AD1*, *AD2*, activation domains), as well as the direct interaction partners. *Square brackets* with the respective amino acid boundaries below the sequences refer to functional regions determined by mutational analysis (*dn*, dominant negative and nuclear receptor-interacting fragments; *HAT*, histone acetyltransferase activity; *p/CAF*, p/CAF interacting fragments; see Voegel et al. 1996, 1998; Oñate et al. 1995, 1997; Kamei et al. 1996; Hayashi et al. 1997; Heery et al. 1997; Spencer et al. 1997; Kalkhoven et al. 1998; Chen et al. 1997). At the *bottom* of the figure, percentage amino acid identities between the three orthologues are denoted for the various regions. Sequence alignment has been performed using the CLUSTALW program (Thompson et al. 1994). GenBank/EMBL accession numbers: *hTIF2*, X97674; *mGRIP1*, U39060; *mNCoA-2*, AF000582; *hSRC-1a*, AJ000881, U90661, U59302, U40396(partial); *hSRC-1e*, AJ000882, U19179(partial); *mSRC-1a* (mNRC1), U64606; *mSRC-1e*, U64828, U56920; *hAIB1*, AF012108; *hTRAM-1*, AF016031; *hACTR*, AF036892; *hRAC3*, AF010227; *mp/CIP*, AF000581

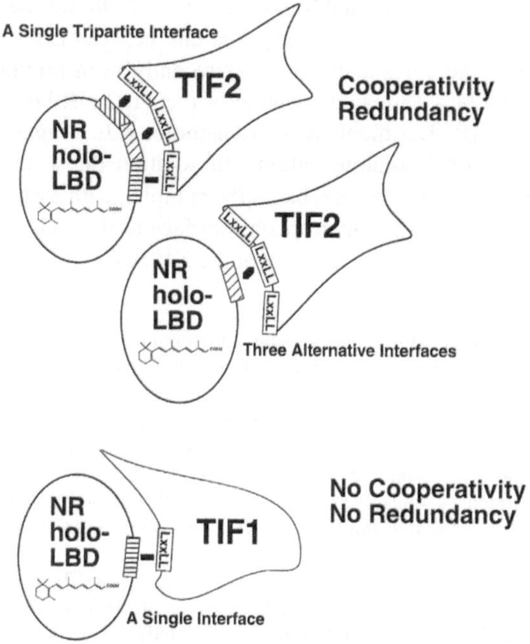

Fig. 3. Possible nuclear receptor–transcriptional intermediary factor (*NR/TIF*) interfaces. TIF2 possesses three NR box motifs ("LxxLL"), thus two different modes of interaction with the NR are conceivable (*top*). In contrast, TIF1 possesses only one NR box motif and only one mode of interaction is conceivable (*bottom*). *NR holo-LBD*, ligand-bound NR ligand-binding domain; *hatched boxes*, hypothetical interaction domain(s) in the LBD, possibly encompassing the activation function 2 activation domain core region

hydrophobic residues from three LBD helices (H3, H4, and H12) which form a shallow hydrophobic cleft as in RARγ and ERα.

Two autonomous activation domains (ADs) have been mapped in TIF2 and SRC-1, one of which (AD1) could not be distinguished in mutagenesis experiments from the CBP interaction domain (CID), while the second (AD2) acts through an as yet unknown mechanism (Kalkhoven et al. 1998; Voegel et al. 1998). To date, no function could be assigned to the most highly conserved N-terminal region, which bears sequence homology with bHLH-PAS proteins (see Sect. 1.3.4).

1.2.4.3 Co-repressors

The antonymous co-repressors have been detected by virtue of their ability to interact in the absence of ligand with NRs and dissociate from the receptor upon ligand binding. The search for such factors was stimulated by the observation that the thyroid hormone (TR) and retinoic acid (RAR) receptors were known to actively repress transcription (Baniahmad et al. 1995 and references therein). Two co-repressors, the nuclear receptor corepressor (N-CoR) (Hörlein et al. 1995) and the silencing mediator for RARs and TRs (SMRT) (Chen and Evans 1995) have been identified to date and functional domains, including two distinct NIDs (not containing LxxLL NR-boxes) and two autonomous repression domains (RDs) have been characterized (Hörlein et al. 1995; Downes et al. 1996; Seol et al. 1996; Zamir et al. 1996; Nagy et al. 1997).

1.2.5 The Emerging General Model of Nuclear Receptor Action

While the identification of NR co-regulators was exciting, it was still not possible to establish a general model because the signal propagation downstream of co-regulators was unknown. However, a burst of recent publications has now provided the missing links, leading to a model with at least (see below) two distinct NR complexes:

1. In the *silencing (or repressing) state*, characterized by the absence of ligand, response element-bound apo-NRs recruit co-repressors (N-CoR, SMRT) that associate with Sin3, which in turn binds to the histone deacetylase (HDAC) RPD3 [Sin3 and RPD3 are recently discovered mammalian homologues of components of the yeast co-repressor complex (Alland et al. 1997; Hassig et al 1997; Heinzel et al. 1997; Kadosh and Struhl 1997; Laherty et al. 1997; Nagy et al. 1997; Zhang et al. 1997; reviewed by Wolffe 1996, 1997; Pazin and Kadonaga 1997)]. The HDAC in this repressor complex is assumed to provoke chromatin condensation as a result of histone deacetylation and the inability to assemble a transcription initiation complex at these sites (Fig. 4a).
2. The *activating state* in response to ligand binding is characterized by the reversal of this process. Response element-bound (holo-)

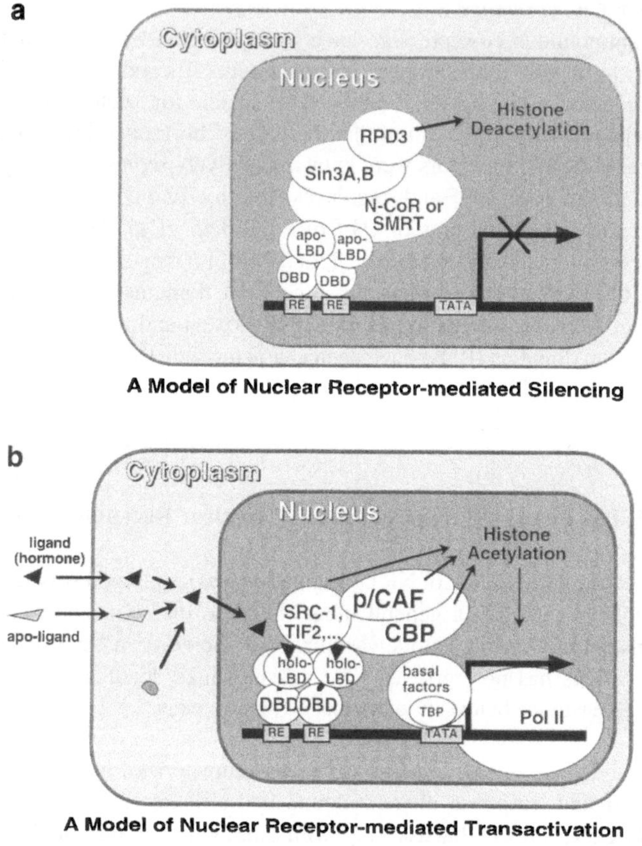

Fig. 4a,b. The proposed models for nuclear receptor (NR) function. A NR homo- or heterodimer targeted to its response element (*RE*) is depicted as two sets of three spheres representing the three functional domains AF-1 (hidden), DNA-binding domain (*DBD*), and ligand-binding domain (*apo-* or *holo-LBD*). **a** Silencing (transrepression) involves the recruitment of corepressors (nuclear receptor corepressor, *N-CoR*; steroid receptor coactivator, *SMRT*) and, indirectly, histone deacetylases like *RPD3*; transcription is blocked, presumably due to chromatin compaction. **b** Transactivation is triggered by binding of the respective ligand (*black triangle*; originating from intra- or extracellular sources), which enables the association of coactivators (e.g., *TIF2*, *SRC-1*, and *CBP*/p300) that possess, or recruit (*p/CAF*), histone acetyltransferase activity. Chromatin is "unfolded", permitting the establishment of a transcription initiation complex

NRs that underwent transconformation recruit co-activators (e.g., TIF2, SRC-1), which in turn bind to histone acetyl transferases (HATs, e.g., CBP/p300, p/CAF; Bannister and Kouzarides 1996; Ogryzko et al. 1996; Yang et al. 1996; Wang et al. 1997); chromatin is consequently "opening up" due to histone acetylation (Fig. 4b). Possibly rather weak direct NR–basal factor interactions represent a second step in the cascade of events leading to transactivation of target genes (see Sect. 1.3.1 below). Surprisingly, several members of the TIF2/SRC-1 family have also been shown to possess HAT activity and recruit p/CAF directly (Chakravarti et al. 1996; Hanstein et al. 1996; Kamei et al. 1996; Yao et al. 1996; Spencer et al. 1997; Chen et al. 1997; for reviews see Eckner 1996; Grunstein 1997; Shikama et al. 1997; Wade et al. 1997; see also Sect. 3.6).

Collectively, the above results indicate that the ligand-modulated interaction between nuclear receptors and co-regulators[4] alters the status of chromatin compaction in the environment of NR binding sites. Note that no statement can be made about the temporal order of events, i.e., currently we do not know whether co-activator–HAT or co-repressor–HDAC complexes are pre-formed prior to recruitment of NRs, nor whether NRs bind co-regulators prior to DNA recognition. This two-state model is of course based on some simplifying assumptions: NRs and all co-factors should be ubiquitously available, NR–DNA interaction is constitutive (see Sect. 1.3.2 below), and solely the availability of ligand is modulable.

While the above concept of acetylation status remodeling is an attractive hypothesis, a significant number of questions have to be addressed and several problems to be solved to generate a comprehensive model of NR-mediated transcriptional repression and activation. Some of the most pertinent questions are pointed out below.

4 Some NRs may not have ligands or AF2 domains and act as repressors only, presumably through interaction with co-repressors; see, for example, Zamir et al. 1996.

1.3 Open Questions

1.3.1 Chromatin Decondensation Is Insufficient for Transactivation

The above-described model considers only chromatin modification as the target of co-regulator function. While it appears to be currently well-accepted that co-regulators have, or recruit, enzymatic activities altering histone acetylation[5], this does not imply that NR-regulated chromatin disruption through histone acetylation is sufficient for transactivation. Indeed, evidence to the contrary has been provided recently (Wong et al. 1997), and the question arises as to how transactivation is induced after the chromatin has been decondensed over a target gene promoter[6]. Two observations may be important in this respect: (a) Members of the TIF2/SRC-1 co-activator family possess two activation domains. One of them (AD1) apparently owes its activity to CBP binding[7], while AD2 acts in a CBP binding-independent manner, perhaps by interacting with other components of the transcriptional machinery; (b) NRs interact not only with co-regulators but also with a number of different components of the transcriptional machinery, such as TFIIB, TBP, TAFs, or TFIIH (Baniahmad et al. 1993; Jacq et al. 1994; Schulman et al. 1995; May et al. 1996; Mengus et al. 1997; Rochette-Egly et al. 1997), albeit in a ligand-independent manner. These interactions may be sufficient to assemble/recruit a transcriptionally competent holo-polymerase II complex, once chromatin decondensation has occurred.

---

5 Even though the final proof demonstrating that the acetylation status of nucleosomal histones in the environment of a NR-regulated enhancer/silencer is indeed altered in vivo upon ligand exposure is still missing.

6 Note that, with respect to the inverse regulatory principle, blocking of HDAC activity by trichostatin A (TSA) is apparently sufficient for relieving transcriptional repression of the *Xenopus* TRβ A gene promoter, but note also that TSA and TR-RXR exhibit distinct requirements for chromatin assembly in transrepression (Wong et al. 1998).

7 This does not solve the problem, however, of AD1 activity in AD2 deletion mutants.

1.3.2 Silencing Requires DNA Binding of apo-NRs

Efficient DNA binding of RXR-RAR heterodimers in vivo, in contrast to their DNA binding in vitro, has been shown to require prior agonist binding (Dey et al. 1994; Chen et al. 1996b). Thus, the question arises as to how a silencing repressor complex can be established over a target gene promoter, since co-repressors apparently bind only to the apo-NR. At least two scenarios would solve this problem: (1) Weak NR–DNA interaction suffices to provide topological information for assembling the complex but cannot be visualized in genomic footprints, and (2) only some promoters can be silenced and bind apo-NRs; the RARβ2 investigated by Dey et al. (1994) and Chen et al. (1996b) belongs to a class of promoters which cannot be silenced.

1.3.3 NR Interaction Pattern of Co-repressors

Albeit that initially only apo-RAR and apo-TR were found to efficiently interact with N-CoR or SMRT[8], and this interaction was shown to correlate with the ability of these apo-NRs to act as transcriptional silencers (Hörlein et al. 1995; Chen and Evans 1995; and references therein), also Rev-Erb, an orphan NR lacking AF-2, apparently acquires his repressor function through co-repressor association (Downes et al. 1996; Zamir et al. 1996). It is more surprising, however, that steroid hormone receptors, not known to contain transcriptional repression domains, interact with co-repressors: In one report N-CoR has been identified as an interacting partner of the RU486-bound progesterone receptor (PR) (Jackson et al. 1997). No interaction was observed with the apo-PR or a PR complexed with the alternative anti-progestin ZK98,299. While these results could be interpreted in the sense that RU486 induces a PR LBD structure which is distinct from its apo-form but happens to provide a surface for co-repressor interaction[9], it is more

[8] Note that apo-RXR was reported to interact only weakly with the co-repressors. However, the role of apo-RXR in co-repressor binding to heterodimers has not yet been determined.

[9] This is possibly an artifactual situation reflecting the overall conservation of NR LBDs.

difficult to envisage how the ER can interact with the SMRT co-repressor, both in the presence and absence of agonists and antagonists (Smith et al. 1997). Future work will have to address the origin of these observations.

1.3.4 Possible Role of the Highly Conserved bHLH and PAS Domains in Co-activators

On the basis of our current understanding of NR function, a cognate DNA binding ability of co-regulators seems superfluous. Thus, it is intriguing that the best conserved region within the TIF2/SRC-1 family of co-activators[10] bears predicted basichelix-loop-helix and Per-Arnt-Sim domains (bHLH-PAS; see Fig. 2). While many bHLH proteins are known as key regulators in myo- and neurogenesis (reviewed in Jan and Jan 1993; Olson and Klein 1994; Yun and Wold 1996; Lee 1997), the members of the bHLH-PAS subfamily are involved in xenobiotic signaling, redox and light sensing, and in circadian oscillation (reviewed in Ponting and Aravind 1997; Sassone-Corsi 1997). Based on crystal structures, the basic region is predicted to bind a DNA half-site, whereas the bHLH-PAS regions form a dimerization interface (Huang et al. 1993; Ma et al. 1994; Wibley et al. 1996).

So far neither direct DNA binding nor dimerization of NR co-activators have been reported. In the absence of experimental data it is tempting to speculate about a possible role of the bHLH-PAS domain. Such speculations comprise the possibility that (a) TIF2/SRC-1 family members may have two separable/independent functions as both NR AF2 co-activators and DNA-binding transactivators for different sets of target genes; (b) non-specific TIF2–DNA binding could modulate transactivation in a gene-specific manner depending on the (additional) presence of a hypothetical TIF2 response element within a NR-responsive promoter; (c) ligand-induced NR transactivation could sequester TIF2 from another signaling pathway [e.g., involving ARNT, which has been

[10] Approximately 60% identity in the C-terminal 350 amino acids between any two of the three, in contrast toapproximately 40% identity in all other regions of similar length. Of the the 240 amino acids of the bHLH-PAS core regions 27% are conserved between TIF2 and AHR or ARNT.

described as a common heterodimerization partner for PAS proteins (Sogawa et al. 1995; Swanson et al. 1995; Probst et al. 1997)]. In this case TIF2 would play the role of an "integrator" between several pathways in a similar way to CBP/p300.

The physiological significance of such potential regulatory scenarios for NR signaling remains obscure. With respect to the above hypothesis (c), Kharat and Saatcioglu (1996) have reported direct mutual transcriptional interference between dioxin and estrogen signaling pathways, and CBP/p300 has been identified as a co-factor for the bHLH-PAS proteins hypoxia-inducible factor-1α and ARNT (Arany et al. 1996; Kobayashi et al. 1997). Therefore, it may be interesting to investigate whether TIF2/SRC-1 family members could play a role in xenobiotic- or oxygen-regulated processes.

1.3.5 Receptor Selectivity and Cell Specificity of Co-activators

At present, for unknown reasons, the androgen receptor (AR) appears to be the only NR that can afford its own co-activators (see Chap. 2, this volume). TIF2/SRC-1 family members[11] are rather promiscuous, although a recent report (Ding et al. 1998) indicates that specific NRs display some co-activator and NR box preferences[12]. Their expression levels, as revealed by Northern blots, vary between tissues but no cell type-specific pattern could be recognized that would distinguish the various co-activators. Thus, it is presently unclear whether, and to what extent, certain co-activators could act in a NR-selective manner. Ongoing gene ablation and future tissue-specific conditional "knock-out" studies in adult animals will help to solve this issue.

11 The same is also true of RIP140, TIF1α, and SUG1.

12 Note, however, that a systematic study addressing the interaction of (several competing) co-regulators with NR homo- or heterodimers bound to their cognate response elements is still missing.

1.3.6 Why So Many HATs?

As discussed above (Sect. 1.2.5), the assembly of co-activator complex(es) involves the accumulation of more than one factor with HAT activity, such as, for example, SRC-1, CBP/p300, and p/CAF. In addition, also $TAF_{II}250$, which can be considered as part of the basal transcriptional machinery, is a HAT (Mizzen et al. 1996). The questions arise whether in cells there are several co-activator complexes, which mediate the activity of multiple NRs (and other activators), and why so many HATs should assemble over target gene promoters. Answering these questions will certainly have to await the characterization of the co-regulator complex(es). It is, however, tempting to speculate that multiple co-activator complexes are involved in gene regulation, since this would introduce an additional level of regulatory complexity: (a) Different classes of transcriptional activators, or even different members of the same superfamily, could interact with (or induce the assembly of) distinct co-activator complexes; (b) several classes of target genes may exist which respond differently to a given level of HAT activity; and (c) it is an attractive possibility that various HATs may have targets distinct from nucleosomal histones, such as, for example, the general transcription factors. Indeed, it was recently shown that $TF_{II}E\beta$ and $TF_{II}F$ are substrates for acetylation by p/CAF and p300 and that considerable specificity exists in substrate utilization by the different acetyltransferases (Imhof et al. 1997). Future work will have to deal with the important issues of defining the identity of (activator-specific) co-activator complexes and the in vivo substrates of the various associated HATs. This will also require tools for the quantification of expression levels of co-regulators in particular cells or tissues.

1.3.7 Are There No AF-1 Co-regulators?

NRs possess two transactivation functions, AF-1 in the N-terminal region A/B and the ligand-dependent AF-2 in the LBD. At present the ligand-dependent interaction of co-regulators has been well documented for AF-2. It is unclear whether AF-1-selective co-regulators exist or whether some of the known AF-2 co-regulators can also interact with AF-1. The existence of AF-1 mediators for the progesterone receptor has been deduced from squelching experiments done by transient transfection (Meyer et al. 1992) and in vitro transcription (Shemshedini et al 1992). It is important to address this point since antagonists may differentially interfere with AF-1 action. For example, in the presence of hydroxy-tamoxifen the ER AF-2 is inactive but AF-1 remains active, while in the presence of ICI164,384 both AF-1 and AF-2 are inactive (Berry et al. 1990). Moreover, AF-1 exhibits cell selectivity (Bocquel et al. 1989; Berry et al. 1990), suggesting the existence of differentially expressed AF-1 mediators. It should be noted that while the interaction of components of the basal transcription machinery with the N-terminal region of NRs has been reported (see Sect. 1.3.1), their significance remains elusive.

1.3.8 Role of Co-regulator Expression and Modification in Disease

One of the recently cloned NR co-activators (AIB1) was identified due to the amplification of its gene in a breast carcinoma (Anzick et al. 1997). Interestingly, we have observed that under certain conditions the presence of overexpressed TIF2 can lead to ligand-independent activity (Voegel et al. 1998)[13]. It should also be noted that increased levels of N-CoR can relieve RAR repression (Söderström et al. 1997). This suggests that transcriptional mediators may possibly be involved in the origin and/or progression of proliferative diseases. Indeed, mutation of the CBP gene has been implicated as the cause of Rubinstein-Taybi syndrome[14] (Petrij et al. 1995) and alterations in the p300 gene were found associated with gastric and colorectal carcinomas (Muraoka et al. 1996). Moreover, the somatic translocation t(8;16) fusing CBP with the acetyltransferase domain of the MOZ gene results in acute myeloid leukemia (Borrow et al. 1996), suggesting that the acetyltransferases p/CAF, CBP, and MOZ may have different target substrates. Also of interest are studies in which the oncogenic capacity of E1 A was com-

[13] Note that these experiments were done with yeast cells, thus excluding any contribution of serum-borne ligands.

[14] Autosomal-dominant disease with a complex phenotype typically including mental retardation, physical abnormalities, and increased incidence of neoplasia. Note that the CBP orthologue p300 can apparently not compensate for these CBP mutations.

promised by co-expressing p300 (Smits et al. 1996). It is thus tempting to speculate that alterations in the cellular co-regulator balance (e.g., overexpression of a co-activator) or altered substrate specificity of the associated enzymatic functions may lead or contribute to pathological states. Therefore, drugs affecting this co-regulator balance or the activities of its individual components in pathological conditions may have therapeutic relevance. In this respect, it will be worthwhile investigating the co-regulator status in anti-hormone-resistant endocrine tumors. In future studies the possible alteration in the activity of co-regulators (see Chap. 5, this volume) due to post-transcriptional modification should also be considered.

1.4 Hierarchies in Heterodimers

All receptors with known ligands contain two autonomous activation functions (AFs), one in the N-terminal region, termed AF-1, which is constitutively active when taken out of the context of the receptor, and a ligand-inducible AF-2 in the LBD. The activity of AF-2 in animal cells is entirely dependent on the integrity of a sequence in the C-terminal part of the LBD, referred to as AF-2 AD core, whose features are conserved among all AF-2-containing receptors (for reviews see Gronemeyer and Laudet 1995; Chambon 1996). Interestingly, the AF-2 of the promiscuous heterodimerization partner RXR presents a unique characteristics, since in RAR or TR heterodimers its activity is not only controlled by its own ligand, but also by the nature and ligand status of its dimerization partner. For example, in heterodimers with non-liganded (apo-) RAR or TR, but not with the orphan NGFI-B, RXR was found incapable of activating transcription in response to its cognate ligand (Forman et al. 1995c; Perlmann and Jansson 1995; Vivat et al. 1997). RXR subordination was most convincingly demonstrated with a constitutively active RXR (Vivat et al. 1997). Although it was initially concluded that RXR was unable to bind its ligand in heterodimers with apo-RAR (Kurokawa et al. 1994), subsequent studies showed very clearly that RXR is fully competent for ligand binding in such heterodimers (Apfel et al. 1995; Chen et al. 1996b, 1998; Kersten et al. 1996; Li et al. 1997).

Ligand autonomy of RAR, TR, and VDR and subordination of RXR in the corresponding heterodimers appears biologically meaningful, since it maintains the identity of a signaling pathway by avoiding initiation/perturbation of thyroid hormone or vitamin D_3 signaling by RXR ligands. Only after its partner is liganded, can RXR synergistically "boost" the response. It will be interesting to understand the structural basis of RXR subordination and the simultaneous ligand autonomy of RXR partners in heterodimeric settings. It remains to be established whether the apparent permissivity for RXR ligand induction of heterodimers with, for example, NGFIB and PPAR does indeed reflect the inability of these apo-receptors to silence RXR, or whether endogenous NGFIB and PPAR ligands present in the cell culture media or the cells themselves have relieved the repression.

1.5 Perspectives for Drug Design

1.5.1 Receptor-Selective and Cell-Specific Ligands

The interspecies conservation of retinoid receptor isoforms, together with the results obtained with isotype-selective retinoids and gene ablation studies, have provided strong evidence that, despite some apparent redundancy in the genetic studies, each of the three RARs has a cognate spectrum of functions (for a review of knockout studies see Kastner et al. 1995). Therefore, given the enormous pharmacological potential of retinoids (Hong and Sporn 1997), several groups have tried to develop isotype-specific agonists and antagonists (for an example of our own group, see Chen et al. 1996b). Interestingly, only three amino acids in the ligand-binding pockets of the three RARs mediate the isotype-specific recognition, as well as the agonistic and antagonistic feature of the synthetic retinoids tested (Gehin et al., in preparation). Currently, an expanding "toolbox" of retinoids exists which display either isotype selectivity (type 1 retinoids, either agonists or antagonists), or mixed agonist/antagonist characteristics for the three RARs (type 2 retinoids) (Gehin et al., in preparation). Some of these retinoids apparently act in a cell-selective manner and it will be exciting to follow the exploitation of these tools in the various cellular models and to define their pharmacological potential.

RXR-specific ligands have also been generated (Lehmann et al. 1992; Boehm et al. 1994; Apfel et al. 1995), which is of particular interest in view of the role of RXR as a promiscuous heterodimerization partner in a number of signaling pathways. Apart from its ability to synergize with RAR ligands [including antagonists (Chen et al. 1996b)], RXR ligands may be used to enhance certain signaling pathways. A recent report, for example, suggests that RXR ligands may stimulate insulin action in non-insulin-dependent diabetes (NIDDM) through the RXR heterodimer with PPARγ and enhance the action of thiazolidinediones (Mukherjee et al. 1997). Of similar importance may be the role of RXR ligands in heterodimers with PPARα, which itself responds to inflammatory mediators (Devchand et al. 1996). Together, the present results suggest that it may be possible, through the design of synthetic ligands, to affect the activity of a RXR-containing NR heterodimer both positively and negatively, and that it may be possible to find "pathway-specific" RXR ligands. Obviously, deciphering the allosteric principles occurring in heterodimers upon interaction with these various RAR and/or RXR ligands will be of great importance for their rational design.

1.5.2 New Receptors for Old Ligands

The identification of a second ER, ERβ, with an expression pattern that is distinct from that of ERα, has fostered hopes that the identification of ER isotype-specific ligands may provide novel estrogens with potentially reduced side effects (for a detailed discussion of ERβ see the chapter by Jan-αke Gustafsson in this issue). This concerns in particular estrogen replacement in post-menopausal women to combat osteoporosis, but a number of other options are being intensively investigated by several groups. Whether isotypes of steroid receptors other than estrogen and progesterone[15] receptors are still to be discovered remains another open question. Particularly exciting, from both a struc-

[15] The human progesterone receptor exists as two isoforms, PR forms A and B, expressed from a single gene but different mRNA families (originating from two clusters of transcriptional start sites). PR form A is identical to form B but lacks the N-terminal 164 residues (for review see Gronemeyer and Laudet 1995).

tural and functional point of view, is the recently discovered pregnane X receptor (Kliewer et al. 1998) (PXR; isoforms PXR.1 and PXR.2) which binds promiscuously to a great number of structurally divergent natural (such as pregnenolone, 17α-hydroxypregnenolone, progesterone, 17α-hydroxyprogesterone, 5β-pregnan-3,20-dione, but not the naturally occurring glucocorticoids, cortisol, and corticosterone) and synthetic (such as dexamethasone, dexamethasone t-butyl acetate, dexamethasone-21-acetate, 6,16α-dimethyl pregnenolone, as well as the glucocorticoid/progestin antagonist RU486 and the anti-glucocorticoid pregnenolone 16α-carbonitrile, PCN) steroid agonists/antagonists and their metabolites. Interestingly, PXR.1-RXR heterodimers activate transcription in response to these various compounds through the DR3 response element (previously thought to be an exclusive response element for VDR-RXR) present in the *CYP3A1* promoter, suggesting that PXR may coordinate steroid and sterol metabolism in liver and intestine.

1.5.3 New Ligands for Old Orphans

The majority of NRs are so-called orphans, which were originally isolated by virtue of their sequence homology with other NRs and no ligand was known (Gronemeyer and Laudet 1995; Mangelsdorf et al. 1995). Several of these orphans have recently found a home, such as the PPARs α and γ which respond to, and interact with, certain prostaglandins and leukotrienes (Forman et al. 1995b; Lehmann et al. 1995; Yu et al. 1995; Kliewer et al. 1997; Devchand et al. 1996), and SF-1 and LXR, which bind oxysterols (Janowski et al. 1996; Lehmann et al. 1997; Lala et al. 1997). Other NRs, like FXR, respond to the name-giving farnesoids (Forman et al. 1995a), but it has remained unclear whether this is the consequence of directly binding the farnesoid or rather a metabolite. Almost certainly, more orphans will find a ligand and this may provide important information about hitherto unrecognized signaling pathways and, thus, new options for drug discovery. But the success with some orphans should not be interpreted to imply that every orphan receptor has a cognate ligand. Indeed, it has been argued that ligand binding is an activity that was acquired during evolution (Escriva et al. 1997). It is therefore possible that a number of "ancient" orphans (for an evolutionary discussion of the NR family see Gronemeyer and Laudet 1995; Laudet 1997) function as constitutive activators or repressors, or are regulated by post-transcriptional modifications. Rev-Erb is an example of a constitutive repressor, since it binds co-repressors but does not possess a functional AF-2 (Adelmant et al. 1996; Downes et al. 1996; Zamir et al. 1996). However, it is tempting to speculate that drug design may take advantage of the structural conservation of the NR family and find ligands even for "true" orphans[16]. Clearly, more information is required about the possible role of the various orphan NRs in pathological settings, but NRs are without doubt an exciting family for drug design and the development of novel approaches in drug discovery.

1.5.4 Exploiting the Signal Transduction Crosstalk

NRs not only act on their cognate gene programs but can also affect, positively or negatively, other signaling pathways, such as those utilizing AP1, NFκB or STAT5. Reciprocally, these pathways can interfere, also positively or negatively, with the activity of NRs, either by directly modifying the receptor, or by altering its ability to transactivate without apparent modification (transcriptional interference). In particular, the signaling crosstalk with AP1 has been shown to be an important target for drug design, but the other options should not be neglected.

1.5.4.1 "Dissociated" Ligands

Previous work has shown that with synthetic ligands it is not only possible to increase isotype selectivity of agonists or antagonists but also to separate transactivation from transrepression of AP1 activity[17]. The signal transduction crosstalk of NRs is not confined to AP1: NRs can also affect Oct-2 A (Wieland et al. 1991), NFκB (Ray and Prefon-

---

[16] Indeed, it cannot be excluded for some of the newly discovered ligands that they are not the cognate ligands but fortuitously fit into the ligand binding pocket. Genetic studies may help to distinguish between the two possibilities.

[17] Such compounds have been termed "dissociated" ligands. For reviews about AP1 transrepression see Pfahl (1993); Beato et al. (1995); and Gronemeyer and Laudet (1995). Note that the crosstalk between NRs and AP1 is (a) reciprocal and (b) can be both positive and negative (Shemshedini et al. 1991; Miner et al. 1991; Bubulya et al. 1996; Paech et al. 1997).

taine 1994; Scheinmann et al. 1995; Brostjan et al. 1996), RelA (Caldenhoven et al. 1995), STAT5 (Stocklin et al. 1996), and Spi-1/PU.1 (Gauthier et al. 1993) activities. The mechanistic basis of these crosstalks is only incompletely understood and it is therefore unclear at present whether ligands can be found which differentially affect the various crosstalk options of a given NR.

Due to the role of AP1 as an "early response gene" AP1 inhibition is of significant importance for pharmacological drug design. Indeed, the anti-inflammatory activity of glucocorticoids may be due to, at least in part, their ability to impair inflammatory cytokine expression[18]. Recently, the first "dissociated" glucocorticoids have been described (Vayssière et al. 1997). This novel class of steroids has the potential to constitute the next generation of glucocorticoids with reduced side effects, which may allow systemic treatment of, for example, rheumatoid arthritis. Importantly, oncogene-transformed cells have been shown to revert to normal phenotypes upon exposure to dissociated retinoids, as they do with their natural ligand retinoic acid (Chen et al. 1995). Identification of the players and pathways involved in these regulatory processes may provide important clues as to the potential of retinoid (and other NR ligand) treatment in proliferative diseases.

1.5.4.2 ERα and ERβ Signal Differently Through AP1 Sites

A spicy addition to the signaling crosstalk theme was the recent recognition that the two ERs, ERα and ERβ, differentially modulate transcription through AP1 sites[19] in response to synthetic estrogens, including the "bone-specific" raloxifen (Paech et al. 1997). Importantly, signaling through cognate estrogen response elements was similar for both receptors and it is tempting to speculate that this signaling option could be exploited to improve the cell selectivity or functionality of estrogens, or exclude unwanted side effects. Clearly, we have to understand both the mechanistic basis of these various crosstalks and the

---

18 Most pro-inflammatory cytokines possess AP1 response elements in their promoters.

19 Ligand-dependent ER signaling through an AP1 site which does not require the integrity of the ER DNA-binding domain was originally reported by Gaub et al. (1990). For an extensive discussion of the ERαβ–AP1 crosstalk see Chap. 7, this volume.

structural features of dissociated ligands, but there is no doubt about the great pharmacological potential in exploring this type of signaling pathway interference.

1.5.4.3 The Next Dimension: Pathways Inducing Receptor and/or TIF Modification

The activities of both NRs (and apparently also of their cognate co-activators (see Chap. 5, this volume) can be modified by post-transcriptional modification. So far only phosphorylations have been reported; examples are the MAP kinase -dependent phosphorylation of the ER (see Chap. 6, this volume) residue Ser118 (Kato et al. 1995; Bunone et al. 1996) or the CDK7-dependent phosphorylation of RARαSer77 (Rochette-Egly et al. 1997). In both these cases the residues are located in the N-terminal region A/B and affect AF-1 activity positively. In contrast, MAP kinase-dependent phosphorylation of PPARγ was reported to inhibit transactivation (Adams et al. 1997). In the ER the stimulation of AF-1 activity may limit the efficiency of antagonist action, which could be relevant, for example, for endocrine therapy of breast cancer.

Whether phosphorylation of particular residues (e.g., ER Tyr537)[20] can generate ligand-independent action of NRs remains to be established. Once the role of receptor and TIF modification in pathophysiology has become clearer, the corresponding enzymes should be re-evaluated as targets for pharmaceutical drug design.

1.5.5 HAT Inhibitors and Interference Drugs

The recognition of histone-modifying enzymes as parts of NR co-activator complex(es)[21] identified a novel drug target. HAT inhibitors may be detected which eradicate the action of individual HATs, since structure–function analyses of the known HATs in the NR co-activator com-

[20] Certain mutations of ERαY537 generate constitutively active receptors (Weis et al. 1996; White et al. 1997) and it has been suggested that phosphorylation may provoke the same effect (see also Chap. 3, this volume).

[21] Note that it is unclear how many distinct co-activator complexes exist in mammalian cells; it is possible that different activators recruit distinct co-activator complexes. Future biochemical and genetic studies may reveal the compositions of cognate co-activator complexes.

plexes (SRC-1, ACTR, CBP/p300, p/CAF, $TAF_{II}250$) revealed very limited sequence similarity in the HAT active fragments. Specific HAT inhibitors may represent a novel type of NR antagonists, which may be useful in the treatment of tumors resistant to classical endocrine therapy.

More speculative is the idea to try and block the interfaces between the various interacting partners in co-activator (or co-repressor) complexes with low molecular weight compounds. Promising techniques for the identification of such small molecule inhibitors of protein–protein interactions in nanodroplets have been established (Huang and Schreiber 1997).

Acknowledgments. We are grateful to the Ernst Schering Research Foundation for generously supporting the workshop which is the basis of the present issue. M.J.S.H. is the recipient of a Marie Curie Fellowship of the European Commission. Work of the group of H.G. that is described here was supported by funds from the Institut National de la Santé et de la Recherche Médicale, the Centre National de la Recherche Scientifique, the Centre Hospitalier Universitaire Régional, the Association pour la Recherche sur le Cancer, the Fondation pour la Recherche Médicale, the Ministère de la Recherche et de la Technologie and the EC BIOMED (contract BMH4-CT96–0181 to H.G.) Programmes, and Bristol-Myers-Squibb.

References

Adams M, Reginato MJ, Shao D, Lazar MA, Chatterjee VK (1997) Transcriptional activation by peroxisome proliferator-activated receptor γ is inhibited by phosphorylation at a consensus mitogen-activated protein kinase site. J Biol Chem 272:5128–5132

Adelmant G, Begue A, Stehelin D, Laudet V (1996) A functional Rev-Erbα responsive element located in the human Rev-Erbα promoter mediates a repressing activity. Proc Natl Acad Sci USA 93:3553–3558

Alland L, Muhle R, Hou H Jr, Potes J, Chin L, Schreiber-Agus N, De Pinho RA (1997) Role for N-CoR and histone deacetylase in Sin3-mediated transcriptional repression. Nature 387:49–55

Anzick SL, Kononen J, Walker RL, Azorsa DO, Tanner MM, Guan XY, Sauter G, Kallioniemi OP, Trent JM, Meltzer PS (1997) AIB1, a steroid receptor coactivator amplified in breast and ovarian cancer. Science 277:965–958

Apfel CM, Kamber M, Klaus M, Mohr P, Keidel S, Le Motte PK (1995) Enhancement of HL-60 differentiation by a new class of retinoids with selective activity on retinoid X receptor. J Biol Chem 270:30765–30772

Arany Z, Huang LE, Eckner R, Bhattacharya S, Jiang C, Goldberg MA, Bunn HF, Livingston DM (1996) An essential role for p300/CBP in the cellular response to hypoxia. Proc Natl Acad Sci USA 93:12969–12973

Baniahmad A, Ha I, Reinberg D, Tsai S, Tsai MJ, O'Malley BW (1993) Interaction of human thyroid hormone receptor β with transcription factor TFIIB may mediate target gene derepression and activation by thyroid hormone. Proc Natl Acad Sci USA 90:8832–8836

Baniahmad A, Leng X, Burris TP, Tsai SY, Tsai MJ, O'Malley BW (1995) The tau 4 activation domain of the thyroid hormone receptor is required for release of a putative corepressor(s) necessary for transcriptional silencing. Mol Cell Biol 15:76–86

Bannister AJ, Kouzarides T (1996) The CBP co-activator is a histone acetyltransferase. Nature 384:641–643

Beato M, Herrlich P, Schütz G (1995) Steroid hormone receptors: many actors in search for a plot. Cell 83:851–857

Berry M, Metzger D, Chambon P (1990) Role of the two activating domains of the oestrogen receptor in the cell-type and promoter-context dependent agonistic activity of the anti-oestrogen 4-hydroxytamoxifen. EMBO J 9:2811–2818

Bocquel MT, Kumar V, Stricker C, Chambon P, Gronemeyer H (1989) The contribution of the N- and C-terminal regions of steroid receptors to activation of transcription is both receptor and cell-specific. Nucleic Acids Res 17:2581–2595

Boehm MF, McClurg MR, Pathirana C, Mangelsdorf D, White SK, Hebert J, Winn D, Goldman ME, Heyman RA (1994) Synthesis of high specific activity [3H]-9-cis-retinoic acid and its application for identifying retinoids with unusual binding properties. J Med Chem 37:408–414

Borrow J, Stanton VP Jr, Andresen JM, Becher R, Behm FG, Chaganti RS, Civin CI, Disteche C, Dube I, Frischauf AM, Horsman D, Mitelman F, Volinia S, Watmore AE, Housman DE (1996) The translocation t(8; 16)(p11; p13) of acute myeloid leukaemia fuses a putative acetyltransferase to the CREB-binding protein. Nat Genet 14:33–41

Bourguet W, Ruff M, Chambon P, Gronemeyer H, Moras D (1995) Crystal structure of the ligand-binding domain of the human nuclear receptor RXRα. Nature 375:377–382

Brostjan C, Anrather J, Csizmadia V, Stroka D, Soares M, Bach FH, Winkler H (1996) Glucocorticoid-mediated repression of NFκB activity in endothelial cells does not involve induction of IκBα synthesis. J Biol Chem 271:19612–19616

Brzozowski AM, Pike AC, Dauter Z, Hubbard RE, Bonn T, Engstrom O, Ohman L, Greene GL, Gustafsson JA, Carlquist M (1997) Molecular basis of agonism and antagonism in the oestrogen receptor. Nature 389:753–758

Bubulya A, Wise SC, Shen XQ, Burmeister LA, Shemshedini L (1996) c-Jun can mediate androgen receptor-induced transactivation. J Biol Chem 271:24583–24589

Bunone G, Briand PA, Miksicek RJ, Picard D (1996) Activation of the unliganded estrogen receptor by EGF involves the MAP kinase pathway and direct phosphorylation. EMBO J 15:2174–2183

Caldenhoven E, Liden J, Wissink S, Van de Stolpe A, Raaijmakers J, Koenderman L, Okret S, Gustafsson JA, Van der Saag PT (1995) Negative cross-talk between RelA and the glucocorticoid receptor: a possible mechanism for the antiinflammatory action of glucocorticoids. Mol Endocrinol 9:401–412

Cavaillès V, Dauvois S, L'Horset F, Lopez G, Hoare S, Kushner PJ, Parker MG (1995) Nuclear factor RIP140 modulates transcriptional activation by the estrogen receptor. EMBO J 14:3741–3751

Chakravarti D, La Morte VJ, Nelson MC, Nakajima T, Schulman IG, Juguilon H, Montminy M, Evans RM (1996) Role of CBP/P300 in nuclear receptor signalling. Nature 383:99–103

Chambon P (1996) A decade of molecular biology of retinoic acid receptors. FASEB J 10:940–954

Chen JD, Evans RM (1995) A transcriptional co-repressor that interacts with nuclear hormone receptors. Nature 377:454–457

Chen JY, Penco S, Ostrowski J, Balaguer P, Pons M, Starrett JE, Reczek P, Chambon P, Gronemeyer H (1995) RAR-specific agonist/antagonists which dissociate transactivation and AP1 transrepression inhibit anchorage-independent cell proliferation. EMBO J 14:1187–1197

Chen JD, Umesono K, Evans RM (1996a) SMRT isoforms mediate repression and anti-repression of nuclear receptor heterodimers. Proc Natl Acad Sci USA 93:7567–7571

Chen JY, Clifford J, Zusi C, Starrett J, Tortolani D, Ostrowski J, Reczek PR, Chambon P, Gronemeyer H (1996b) Two distinct actions of retinoid-receptor ligands. Nature 382:819–822

Chen H, Lin RJ, Schiltz RL, Chakravarti D, Nash A, Nagy L, Privalsky ML, Nakatani Y, Evans RM (1997) Nuclear receptor coactivator ACTR is a novel histone acetyltransferase and forms a multimeric activation complex with p/CAF and CBP/p300. Cell 90:569–580

Chen ZP, Iyer J, Bourguet W, Held P, Mioskowski C, Lebeau L, Noy N, Chambon P, Gronemeyer H (1998) Ligand- and DNA-induced dissociation of RXR tetramers. J Mol Biol 275:55–65

De Groot LJ (1995) Endocrinology. Saunders, Philadelphia

Devchand PR, Keller H, Peters JM, Vazquez M, Gonzalez FJ, Wahli W (1996) The PPARα-leukotriene B4 pathway to inflammation control. Nature 384:39–43

Dey A, Minucci S, Ozato K (1994) Ligand-dependent occupancy of the retinoic acid receptor β2 promoter in vivo. Mol Cell Biol 14:8191–8201

Ding XF, Anderson CM, Ma H, Hong H, Uht RM, Kushner PJ, Stallcup MR (1998) Nuclear receptor binding sites of coactivators GRIPI and SRC-1: multiple motifs with different binding specificities. Mol Endocrinol 12:302–313

Downes M, Burke LJ, Bailey PJ, Muscat GE (1996) Two receptor interaction domains in the corepressor, N-CoR/RIP13, are required for an efficient interaction with Rev-erbAα and RVR: physical association is dependent on the E region of the orphan receptors. Nucleic Acids Res 24:4379–4386

Eckner R (1996) p300 and CBP as transcriptional regulators and targets of oncogenic events. Biol Chem 377:685–688

Eggert M, Mows CC, Tripier D, Arnold R, Michel J, Nickel J, Schmidt S, Beato M, Renkawitz R (1995) A fraction enriched in a novel glucocorticoid receptor-interacting protein stimulates receptor-dependent transcription in vitro. J Biol Chem 270:30755–30759

Enmark E, Gustafsson Jα (1996) Orphan nuclear receptors – the first eight years. Mol Endocrinol 10:1293–1307

Escriva H, Safi R, Hanni C, Langlois MC, Saumitou-Laprade P, Stehelin D, Capron A, Pierce R, Laudet V (1997) Ligand binding was acquired during evolution of nuclear receptors. Proc Natl Acad Sci USA 94:6803–6808

Forman BM, Goode E, Chen J, Oro AE, Bradley DJ, Perlmann T, Noonan DJ, Burka LT, McMorris T, Lamph WW, Evans RM, Weinberger C (1995a) Identification of a nuclear receptor that is activated by farnesol metabolites. Cell 81:687–693

Forman BM, Tontonoz P, Chen J, Brun RP, Spiegelman BM, Evans RM (1995b) 15-Deoxy-$\Delta^{12,14}$-prostaglandin J2 is a ligand for the adipocyte determination factor PPARγ. Cell 83:803–812

Forman BM, Umesono K, Chen J, Evans RM (1995c) Unique response pathways are established by allosteric interactions among nuclear hormone receptors. Cell 81:541–550

Gaub MP, Bellard M, Scheuer I, Chambon P, Sassone-Corsi P (1990) Activation of the ovalbumin gene by the estrogen receptor involves the fos-jun complex. Cell 63:1267–1276

Gauthier JM, Bourachot B, Doucas V, Yaniv M, Moreau-Gachelin F (1993) Functional interference between the Spi-1/PU.1 oncoprotein and steroid hormone or vitamin receptors. EMBO J 12:5089–5096

Glass CK, Rose DW, Rosenfeld MG (1997) Nuclear receptor coactivators. Curr Opin Cell Biol 9:222–232

Gronemeyer H, Laudet V (1995) Transcription factors 3: nuclear receptors. Protein Profile 2:1173–1308

Grunstein M (1997) Histone acetylation in chromatin structure and transcription. Nature 389:349–352

Hanstein B, Eckner R, Di Renzo J, Halachmi S, Liu H, Searcy B, Kurokawa R, Brown M (1996) p300 is a component of an estrogen receptor coactivator complex. Proc Natl Acad Sci USA 93:11540–11545

Hassig CA, Fleischer TC, Billin AN, Schreiber SL, Ayer DE (1997) Histone deacetylase activity is required for full transcriptional repression by mSin3 A. Cell 89:341–347

Hayashi Y, Ohmori S, Ito T, Seo H (1997) A splicing variant of steroid receptor coactivator-1 (SRC-1E): the major isoform of SRC-1 to mediate thyroid hormone action. Biochem Biophys Res Commun 236:83–87

Heery DM, Kalkhoven E, Hoare S, Parker MG (1997) A signature motif in transcriptional co-activators mediates binding to nuclear receptors. Nature 387:733–736

Heinzel T, Lavinsky RM, Mullen TM, Soderstrom M, Laherty CD, Torchia J, Yang WM, Brard G, Ngo SD, Davie JR, Seto E, Eisenman RN, Rose DW, Glass CK, Rosenfeld MG (1997) A complex containing N-CoR, mSin3 and histone deacetylase mediates transcriptional repression. Nature 387:43–48

Hong H, Kohli K, Trivedi A, Johnson DL, Stallcup MR (1996) GRIP1, a novel mouse protein that serves as a transcriptional coactivator in yeast for the hormone binding domains of steroid receptors. Proc Natl Acad Sci USA 93:4948–4952

Hong H, Kohli K, Garabedian MJ, Stallcup MR (1997) GRIP1, a transcriptional coactivator for the AF-2 transactivation domain of steroid, thyroid, retinoid, and vitamin D receptors. Mol Cell Biol 17:2735–2744

Hong WK, Sporn MB (1997) Recent advances in chemoprevention of cancer. Sience 278:1073–1077

Hörlein AJ, Näär AM, Heinzel T, Torchia J, Gloss B, Kurokawa R, Ryan A, Kamei Y, Söderström M, Glass CK, Rosenfeld MG (1995) Ligand-independent repression by the thyroid hormone receptor mediated by a nuclear receptor co-repressor. Nature 377:397–404

Horwitz KB, Jackson TA, Bain DL, Richer JK, Takimoto GS, Tung L (1996) Nuclear receptor coactivators and corepressors. Mol Endocrinol 10:1167–1177

Huang J, Schreiber SL (1997) A yeast genetic system for selecting small molecule inhibitors of protein-protein interaction in nanodroplets. Proc Natl Acad Sci USA 94:13396–13401

Huang ZJ, Edery I, Rosbash M (1993) PAS is a dimerization domain common to Drosophila period and several transcription factors. Nature 364:259–262

Imhof A, Yang XJ, Ogryzko VV, Nakatani Y, Wolffe AP, Ge H (1997) Acetylation of general transcription factors by histone acetyltransferases. Curr Biol 7:689–692

Jackson TA, Richer JK, Bain DL, Takimoto GS, Tung L, Horwitz KB (1997) The partial agonist activity of antagonist-occupied steroid receptors is controlled by a novel hinge domain-binding coactivator L7/SPA and the corepressors N-CoR or SMRT. Mol Endocrinol 11:693–705

Jacq X, Brou C, Lutz Y, Davidson I, Chambon P, Tora L (1994) Human $TAF_{II}30$ is present in a distinct TFIID complex and is required for transcriptional activation by the estrogen receptor. Cell 79:107–117

Jan YN, Jan LY (1993) HLH proteins, fly neurogenesis, and vertebrate myogenesis. Cell 75:827–830

Janowski BA, Willy PJ, Rama Devi T, Falck JR, Mangelsdorf DJ (1996) An oxysterol signalling pathway mediated by the nuclear receptor LXRα. Nature 383:728–731

Kadosh D, Struhl K (1997) Repression by Ume6 involves recruitment of a complex containing Sin3 corepressor and Rpd3 histone deacetylase to target promoters. Cell 89:365–371

Kalkhoven E, Valentine JE, Heery DM, Parker MG (1998) Isoforms of steroid receptor coactivator 1 differ in their ability to potentiate transcription by the oestrogen receptor. EMBO J 17:232–243

Kamei Y, Xu L, Heinzel T, Torchia J, Kurokawa R, Gloss B, Lin SC, Heyman RA, Rose DW, Glass CK, Rosenfeld MG (1996) A CBP integrator complex mediates transcriptional activation and AP-1 inhibition by nuclear receptors. Cell 85:403–414

Kastner P, Mark M, Chambon P (1995) Nonsteroid nuclear receptors: what are genetic studies telling us about their role in real life. Cell 83:859–869

Kato S, Endoh H, Masuhiro Y, Kitamoto T, Uchiyama S, Sasaki H, Masushige S, Gotoh Y, Nishida E, Kawashima H, Metzger D, Chambon P (1995) Activation of the estrogen receptor through phosphorylation by mitogen-activated protein kinase. Science 270:1491–1494

Kersten S, Dawson MI, Lewis BA, Noy N (1996) Individual subunits of heterodimers comprised of retinoic acid and retinoid X receptors interact with their ligands independently. Biochemistry 35:3816–3824

Kharat I, Saatcioglu F (1996) Antiestrogenic effects of 2,3,7,8-tetrachlorodibenzo-p-dioxin are mediated by direct transcriptional interference with the liganded estrogen receptor. J Biol Chem 271:10533–10537

Kliewer SA, Sundseth SS, Jones SA, Brown PJ, Wisely GB, Koble CS, Devchand P, Wahli W, Willson TM, Lenhard JM, Lehmann JM (1997) Fatty acids and eicosanoids regulate gene expression through direct interactions with peroxisome proliferator-activated receptors α and γ5D. Proc Natl Acad Sci USA 94:4318–4323

Kliewer SA, Moore JT, Wade L, Staudinger JL, Watson MA, Jones SA, McKee DD, Oliver BB, Willson TM, Zetterström MH, Perlmann T, Lehmann JM (1998) An orphan nuclear receptor activated by pregnanes defines a novel steroid signaling pathway. Cell 92 (in press)

Kobayashi A, Numayama-Tsuruta K, Sogawa K, Fujii-Kuriyama Y (1997) CBP/p300 functions as a possible transcriptional coactivator of Ah receptor nuclear translocator (Arnt). J Biochem 122:703–710

Korach KS, Couse JF, Curtis SW, Washburn TF, Lindzey J, Kimbro KS, Eddy EM, Migliaccio S, Snedeker SM, Lubahn DB, Schomberg DW, Smith EP (1996) Estrogen receptor gene disruption: molecular characterization and experimental and clinical phenotypes. Rec Prog Horm Res 51:159–158

Krey G, Braissant O, L'Horset F, Kalkhoven E, Perroud M, Parker MG, Wahli W (1997) Fatty acids, eicosanoids, and hypolipidemic agents identified as ligands of peroxisome proliferator-activated receptors by coactivator-dependent receptor ligand assay. Mol Endocrinol 11:779–791

Kuiper GGJM, Enmark E, Pelto-Huikko M, Nilsson S, Gustafsson JA (1996) Cloning of a novel estrogen receptor expressed in rat prostate and ovary. Proc Natl Acad Sci USA 93:5925–5930

Kurokawa R, Di Renzo J, Boehm M, Sugarman J, Gloss B, Rosenfeld MG, Heyman RA, Glass CK (1994) Regulation of retinoid signalling by receptor polarity and allosteric control of ligand binding. Nature 371:528–531

Laherty CD, Yang WM, Sun JM, Davie JR, Seto E, Eisenman RN (1997) Histone deacetylases associated with the mSin3 corepressor mediate mad transcriptional repression. Cell 89:349–356

Lala DS, Syka PM, Lazarchik SB, Mangelsdorf DJ, Parker KL, Heyman RA (1997) Activation of the orphan nuclear receptor steroidogenic factor 1 by oxysterols. Proc Natl Acad Sci USA 94:4895–4900

Laudet V (1997) Evolution of the nuclear receptor superfamily: early diversification from an ancestral orphan receptor. J Mol Endocrinol 19:207–226

Le Douarin B, Pierrat B, vom Baur E, Chambon P, Losson R (1995) A new version of the two-hybrid assay for detection of protein-protein interactions. Nucleic Acids Res 23:876–878

Le Douarin B, Nielsen AL, Garnier JM, Ichinose H, Jeanmougin F, Losson R, Chambon P (1996) A possible involvement of TIF1α and TIF1β in the epigenetic control of transcription by nuclear receptors. EMBO J 15:6701–6715

Lee JE (1997) Basic helix-loop-helix genes in neural development. Curr Opin Neurobiol 7:13–20

Lee JW, Ryan F, Swaffield JC, Johnston SA, Moore DD (1995) Interaction of thyroid-hormone receptor with a conserved transcriptional mediator. Nature 374:91–94

Lehmann JM, Jong L, Fanjul A, Cameron JF, Lu XP, Haefner P, Dawson MI, Pfahl M (1992) Retinoids selective for retinoid X receptor response pathways. Science 258:1944–1946

Lehmann JM, Moore LB, Smith-Oliver TA, Wilkison WO, Willson TM, Kliewer SA (1995) An antidiabetic thiazolidinedione is a high affinity ligand for peroxisome proliferator-activated receptor γ (PPARγ). J Biol Chem 270:12953–12956

Lehmann JM, Kliewer SA, Moore LB, Smith-Oliver TA, Oliver BB, Su JL, Sundseth SS, Winegar DA, Blanchard DE, Spencer TA, Willson TM (1997) Activation of the nuclear receptor LXR by oxysterols defines a new hormone response pathway. J Biol Chem 272:3137–3140

Li H, Gomes PJ, Don Chen J (1997) RAC3, a steroid/nuclear receptor-associated coactivator that is related to SRC-1 and TIF2. Proc Natl Acad Sci USA 94:8479–8484

Lydon JP, De Mayo FJ, Funk CR, Mani SK, Hughes AR, Montgomery CA Jr, Shyamala G, Conneely OM, O'Malley BW (1995) Mice lacking progesterone receptor exhibit pleiotropic reproductive abnormalities. Genes Dev 9:2266–2278

Lyon J, La Thangue NB (1997) Chromatin research gathers pace. Trends Cell Biol 7:389

Ma PC, Rould MA, Weintraub H, Pabo CO (1994) Crystal structure of MyoD bHLH domain-DNA complex: perspectives on DNA recognition and implications for transcriptional activation. Cell 77:451–459

Mangelsdorf DJ, Evans RM (1995) The RXR heterodimers and orphan receptors. Cell 83:841–850

Mangelsdorf DJ, Thummel C, Beato M, Herrlich P, Schütz G, Umesono K, Blumberg B, Kastner P, Mark M, Chambon P, Evans RM (1995) The nuclear receptor superfamily: the second decade. Cell 83:835–839

May M, Mengus G, Lavigne AC, Chambon P, Davidson I (1996) Human TAF(II)28 promotes transcriptional stimulation by activation function 2 of the retinoid X receptors. EMBO J 15:3093–3104

Mengus G, May M, Carre L, Chambon P, Davidson I (1997) Human TAF(II)135 potentiates transcriptional activation by the AF-2 s of the retinoic acid, vitamin D3, and thyroid hormone receptors in mammalian cells. Genes Dev 11:1381–1395

Meyer ME, Gronemeyer H, Turcotte B, Bocquel MT, Tasset D, Chambon P (1989) Steroid hormone receptors compete for factors that mediate their enhancer function. Cell 57:433–442

Meyer ME, Quirin-Stricker C, Lerouge T, Bocquel MT, Gronemeyer H (1992) A limiting factor mediates the differential activation of promoters by the human progesterone receptor isoforms. J Biol Chem 267:10882–10887

Milburn MV, Charifson P, Lambert M, Cobb J, Wisely GB (1997) Ligand binding and activation of PPARs – crystal structures of PPARγ. Abstract of the EMBO workshop on "structure and function of nuclear receptors", 2–5 May 1997, Erice, Italy.

Miner JN, Diamond MI, Yamamoto KR (1991) Joints in the regulatory lattice: composite regulation by steroid receptor-AP1 complexes. Cell Growth Differ 2:525–530

Mizzen CA, Yang XJ, Kokubo T, Brownell JE, Bannister AJ, Owen-Hugues T, Workman J, Wang L, Berger SL, Kouzarides T, Nakatani Y, Allis CD (1996) The $TAF_{II}250$ subunit of TFIID has histone acetyltransferase activity. Cell 87:1261–1270

Mukherjee R, Davies PJ, Crombie DL, Bischoff ED, Cesario RM, Jow L, Hamann LG, Boehm MF, Mondon CE, Nadzan AM, Paterniti JR Jr, Heyman RA (1997) Sensitization of diabetic and obese mice to insulin by retinoid X receptor agonists. Nature 386:407–410

Muraoka M, Konishi M, Kikuchi-Yanoshita R, Tanaka K, Shitara N, Chong JM, Iwama T, Miyaki M (1996) p300 gene alterations in colorectal and gastric carcinomas. Oncogene 12:1565–1569

Nagy L, Kao HY, Chakravarti D, Lin RJ, Hassig CA, Ayer DE, Schreiber SL, Evans RM (1997) Nuclear receptor repression mediated by a complex containing SMRT, mSin3 A, and histone deacetylase. Cell 89:373–380

Ogryzko VV, Schiltz RL, Russanova V, Howard BH, Nakatani Y (1996) The transcriptional coactivators p300 and CBP are histone acetyltransferases. Cell 87:953–959

Olson EN, Klein WH (1994) bHLH factors in muscle development: dead lines and commitments, what to leave in and what to leave out. Genes Dev 8:1–8

Oñate SA, Tsai SY, Tsai MJ, O'Malley BW (1995) Sequence and characterization of a coactivator for the steroid hormone receptor superfamily. Science 270:1354–1357

Paech K, Webb P, Kuiper GG, Nilsson S, Gustafsson J, Kushner PJ, Scanlan TS (1997) Differential ligand activation of estrogen receptors ERα and ERβ at AP1 sites. Science 277:1508–1510

Pazin MJ, Kadonaga JT (1997) What's up and down with histone deacetylation and transcription? Cell 89:325–328

Perlmann T, Jansson L (1995) A novel pathway for vitamin A signaling mediated by RXR heterodimerization with NGFI-B and NURR1. Genes Dev 9:769–782

Petrij F, Giles RH, Dauwerse HG, Saris JJ, Hennekam RC, Masuno M, Tommerup N, van Ommen GJ, Goodman RH, Peters DJM, Breuning MH (1995) Rubinstein-Taybi syndrome caused by mutations in the transcriptional co-activator CBP. Nature 376:348–351

Pfahl M (1993) Nuclear receptor/AP-1 interaction. Endocr Rev 14:651–658

Ponting C, Aravind L (1997) PAS, a multifunctional domain family comes to light. Curr Biol 7:R674–R677

Probst MR, Fan CM, Tessier-Lavigne M, Hankinson O (1997) Two murine homologs of the Drosophila single-minded protein that interact with the mouse aryl hydrocarbon receptor nuclear translocator protein. J Biol Chem 272:4451–4457

Ray A, Prefontaine KE (1994) Physical association and functional antagonism between the p65 subunit of transcription factor NFκ and the glucocorticoid receptor. Proc Natl Acad Sci USA 91:752–756

Renaud JP, Rochel N, Ruff M, Vivat V, Chambon P, Gronemeyer H, Moras D (1995) Crystal structure of the RARγ ligand-binding domain bound to all-trans retinoic acid. Nature 378:681–689

Rochette-Egly C, Adam S, Rossignol M, Egly JM, Chambon P (1997) Stimulation of RARα activation function AF-1 through binding to the general transcription factor TFIIH and phosphorylation by CDK7. Cell 90:97–107

Sassone-Corsi P (1997) Molecular clocks. PERpetuating the PASt [news; comment]. Nature 389:443–444

Scheinman RI, Gualberto A, Jewell CM, Cidlowski JA, Baldwin AS Jr (1995) Characterization of mechanisms involved in transrepression of NFkb by activated glucocorticoid receptors. Mol Cell Biol 15:943–953

Schulman IG, Chakravarti D, Juguilon H, Romo A, Evans RM (1995) Interactions between the retinoid X receptor and a conserved region of the TATA-binding protein mediate hormone-dependent transactivation. Proc Natl Acad Sci USA 92:8288–8292

Seol W, Mahon MJ, Lee YK, Moore DD (1996) Two receptor interacting domains in the nuclear hormone receptor corepressor RIP13/N-CoR. Mol Endocrinol 10:1646–1655

Shemshedini L, Knauthe R, Sassone-Corsi P, Pornon A, Gronemeyer H (1991) Cell-specific inhibitory and stimulatory effects of Fos and Jun on transcription activation by nuclear receptors. EMBO J 10:3839–3849

Shemshedini L, Ji JW, Brou C, Chambon P, Gronemeyer H (1992) In vitro activity of the transcription activation functions of the progesterone receptor. Evidence for intermediary factors. J Biol Chem 267:1834–1839

Shikama N, Lyon J, La Thangue NB (1997) The p300/CBP family: integrating signals with transcription factors and chromatin. Trends Cell Biol 7:230–236

Smith CL, Nawaz Z, O'Malley BW (1997) Coactivator and corepressor regulation of the agonist/antagonist activity of the mixed antiestrogen, 4-hydroxytamoxifen. Mol Endocrinol 11:657–666

Smits PH, de Wit L, van der Eb AJ, Zantema A (1996) The adenovirus E1A-associated 300 kDa adaptor protein counteracts the inhibition of the collagenase promoter by E1 A and represses transformation. Oncogene 12:1529–1535

Söderström M, Vo A, Heinzel T, Lavinsky RM, Yang WM, Seto E, Peterson DA, Rosenfeld MG, Glass CK (1997) Differential effects of nuclear receptor corepressor (N-CoR) expression levels on retinoic acid receptor-mediated repression support the existence of dynamically regulated corepressor complexes. Mol Endocrinol 11:682–692

Sogawa K, Nakano R, Kobayashi A, Kikuchi Y, Ohe N, Matsushita N, Fujii-Kuriyama Y (1995) Possible function of Ah receptor nuclear translocator (Arnt) homodimer in transcriptional regulation. Proc Natl Acad Sci USA 92:1936–1940

Spencer TE, Jenster G, Burcin MM, Allis CD, Zhou J, Mizzen CA, McKenna NJ, Oñate SA, Tsai SY, Tsai MJ, O'Malley BW (1997) Steroid receptor coactivator-1 is a histone acetyltransferase. Nature 389:194–198

Sporn MB, Roberts AB, Goodman DS (1994) The retinoids. Biology, chemistry and medicine. Raven, New York

Stocklin E, Wissler M, Gouilleux F, Groner B (1996) Functional interactions between Stat5 and the glucocorticoid receptor. Nature 383:726–728

Swanson HI, Chan WK, Bradfield CA (1995) DNA binding specificities and pairing rules of the Ah receptor, ARNT, and SIM proteins. J Biol Chem 270:26292–26302

Takeshita A, Cardona GR, Koibuchi N, Suen CS, Chin WW (1997) TRAM-1, a novel 160-kDa thyroid hormone receptor activator molecule, exhibits distinct properties from steroid coactivator-1. J Biol Chem 272:27629–27634

Thompson JD, Higgins DG, Gibson TJ (1994) CLUSTAL W: improving the sensitivity of progressive multiple sequence alignment through sequence weighting, positions-specific gap penalties and weight matrix choice. Nucleic Acids Res 22:4673–4680

Torchia J, Rose DW, Inostroza J, Kamei Y, Westin S, Glass CK, Rosenfeld MG (1997) The transcriptional co-activator p/CIP binds CBP and mediates nuclear-receptor function. Nature 387:677–684

Tremblay GB, Tremblay A, Copeland NG, Gilbert DJ, Jenkins NA, Labrie F, Giguere V (1997) Cloning, chromosomal localization, and functional analysis of the murine estrogen receptor β. Mol Endocrinol 11:353–365

Vayssière BM, Dupont S, Choquart A, Petit F, Garcia T, Marchandeau C, Gronemeyer H, Resche-Rigon M (1997) Synthetic glucocorticoids that dissociate transactivation and AP-1 transrepression exhibit antiinflammatory activity in vivo. Mol Endocrinol 11:1245–1255

Vivat V, Zechel C, Wurtz JM, Bourguet W, Kagechika H, Umemiya H, Shudo K, Moras D, Gronemeyer H, Chambon P (1997) A mutation mimicking ligand-induced conformational change yields a constitutive RXR that senses allosteric effects in heterodimers. EMBO J 16:5697–5709

Voegel JJ, Heine MJS, Zechel C, Chambon P, Gronemeyer H (1996) TIF2, a 160 kDa transcriptional mediator for the ligand-dependent activation function AF-2 of nuclear receptors. EMBO J 15:3667–3675

Voegel JJ, Heine MJS, Tini M, Vivat V, Chambon P, Gronemeyer H (1998) The coactivator TIF2 contains three nuclear receptor-binding motifs and mediates transactivation through CBP binding-dependent and -independent pathways. EMBO J 17:507–519

Vom Baur E, Zechel C, Heery D, Heine MJ, Garnier JM, Vivat V, Le Douarin B, Gronemeyer H, Chambon P, Losson R (1996) Differential ligand-dependent interactions between the AF-2 activating domain of nuclear receptors and the putative transcriptional intermediary factors mSUG1 and TIF1. EMBO J 15:110–124

Wade PA, Pruss D, Wolffe AP (1997) Histone acetylation: chromatin in action. Trends Biochem Sci 22:128–132

Wagner RL, Apriletti JW, McGrath ME, West BL, Baxter JD, Fletterick RJ (1995) A structural role for hormone in the thyroid hormone receptor. Nature 378:690–697

Wang L, Mizzen C, Ying C, Candau R, Barlev N, Brownell J, Allis CD, Berger SL (1997) Histone acetyltransferase activity is conserved between yeast and human GCN5 and is required for complementation of growth and transcriptional activation. Mol Cell Biol 17:519–527

Warrell RP Jr, Frankel SR, Miller WH Jr, Scheinberg DA, Itri LM, Hittelman WN, Vyas R, Andreeff M, Tafuri A, Jakubowski A (1991) Differentiation therapy of acute promyelocytic leukemia with tretinoin (all-trans-retinoic acid). N Engl J Med 324:1385–1393

Weis KE, Ekena K, Thomas JA, Lazennec G, Katzenellenbogen BS (1996) Constitutively active human estrogen receptors containing amino acid substitutions for tyrosine 537 in the receptor protein. Mol Endocrinol 10:1388–1398

White R, Sjoberg M, Kalkhoven E, Parker MG (1997) Ligand-independent activation of the oestrogen receptor by mutation of a conserved tyrosine. EMBO J 16:1427–1435

Wibley J, Deed R, Jasiok M, Douglas K, Norton J (1996) A homology model of the Id-3 helix-loop-helix domain as a basis for structure-function predictions. Biochim Biophys Acta 1294:138–146

Wieland S, Dobbeling U, Rusconi S (1991) Interference and synergism of glucocorticoid receptor and octamer factors. EMBO J 10:2513–2521

Wolffe AP (1996) Histone deacetylase: a regulator of transcription [comment]. Science 272:371–372

Wolffe AP (1997) Transcriptional control. Sinful repression [news; comment]. Nature 387:16–17

Wong J, Shi YB, Wolffe AP (1997) Determinants of chromatin disruption and transcriptional regulation instigated by the thyroid hormone receptor: hormone-regulated chromatin disruption is not sufficient for transcriptional activation. EMBO J 16:3158–3171

Wong J, Patterton D, Imhof A, Guschin D, Shi Y-B, Wolffe AP (1998) Distinct requirements for chromatin assembly in transcriptional repression by thyroid hormone receptor and histone deacetylase. EMBO J 17 (in press)

Wurtz JM, Bourguet W, Renaud JP, Vivat V, Chambon P, Moras D, Gronemeyer H (1996) A canonical structure for the ligand-binding domain of nuclear receptors [published erratum appears in Nat Struct Biol 1996 Feb; 3(2):206]. Nat Struct Biol 3:87–94

Yang XJ, Ogryzko VV, Nishikawa J-I, Howard BH, Nakatani Y (1996) A p300/CBP-associated factor that competes with the adenoviral oncoprotein El A. Nature 382:319–324

Yao TP, Ku G, Zhou N, Scully R, Livingston DM (1996) The nuclear hormone receptor coactivator SRC-1 is a specific target of p300. Proc Natl Acad Sci USA 93:10626–10631

Yu K, Bayona W, Kallen CB, Harding HP, Ravera CP McMahon G, Brown M, Lazar MA (1995) Differential activation of peroxisome proliferator-activated receptors by eicosanoids. J Biol Chem. 270:23975–23985

Yun K, Wold B (1996) Skeletal muscle determination and differentiation: story of a core regulatory network and its context. Curr Opin Cell Biol 8:877–889

Zamir I, Harding HP, Atkins GB, Horlein A, Glass CK, Rosenfeld MG, Lazar MA (1996) A nuclear hormone receptor corepressor mediates transcriptional silencing by receptors with distinct repression domains. Mol Cell Biol 16:5458–5465

Zeiner M, Gehring U (1995) A protein that interacts with members of the nuclear hormone receptor family: identification and cDNA cloning. Proc Natl Acad Sci USA 92:11465–11469

Zhang Y, Iratni R, Erdjument-Bromage H, Tempst P, Reinberg D (1997) Histone deacetylases and SAP18, a novel polypeptide, are components of a human Sin3 complex. Cell 89:357–364

2 The Androgen Receptor Co-regulator, ARA_{70}

H. Miyamoto, S. Yeh, S.-B. Hwang, N. Fujimoto, P. Hsiao, H. Ting, K. Nishimura, C. Wang, S. Inui, H. Uemura, H. Kang, and C. Chang

2.1 Cloning of ARA_{70}

Androgens play an important role in the process of male sexual differentiation and development. Their actions are mediated by the androgen receptor (AR), which is a member of a large family of ligand-dependent transregulators known as the steroid receptor superfamily (Chang et al. 1988; Evans 1988; Lubahn et al. 1988). To further understand the mechanism of androgen-AR action, we have applied a yeast two-hybrid system to identify the AR-associated proteins. An AR-associated protein, ARA_{70}, was isolated from a brain cDNA library using the GAL4 DNA binding domain (DBD) fused with human AR peptide (amino acids 595–918) as bait (Yeh and Chang 1996). The β-galactosidase liquid assay in the yeast showed that ARA_{70} interacted with AR, but not with other nuclear receptors, such as retinoic acid receptor (RXR) and TR4 orphan receptor.

To test whether ARA_{70} can affect the transcriptional activity of AR, AR and ARA_{70} were co-transfected into human prostate cancer DU145 cells under eukaryotic promoter control. Ligand-free AR has a minimal reporter gene activity of mouse mammary tumor virus-androgen response element-chloramphenicol acethyltransferase (MMTV-ARE-CAT) with or without co-transfection of ARA_{70}. Addition of dihydrotestosterone (DHT) results in a between five and six-fold increase of AR activity. Furthermore, this transcriptional activity can be increased 40- to 60-fold by co-transfection of ARA_{70} with 1.5 μ g of AR in a dose-dependent manner, reaching a plateau at 4.5 μ g of ARA_{70} in DU145 cells. To rule out any indirect effects on the basal activity of the MMTV-ARE CAT reporter, the ARE DNA fragment was removed from the reporter plasmid. The results showed that ARA_{70} induced no activity on this reporter in the presence or absence of androgen. These data suggested that stimulation of AR transcriptional activity by ARA_{70} may occur through a ligand-bound AR. On the other hand, the transcriptional activity of other steroid receptors, such as the glucocorticoid receptor, progesterone receptor, and estrogen receptor cannot be induced significantly (near two-fold) by ARA_{70} in the presence of their own ligands. These findings clearly indicated that ARA_{70} is the first identified ligand-dependent associated protein that might function as a relatively specific co-activator for AR.

To further confirm the interaction between AR and ARA_{70}, we then applied an in vitro immunoprecipitation assay with an AR antibody (CW2). We demonstrated that CW2 can co-precipitate AR and ARA_{70} when incubated with *in vitro* transcribed/translated full-length human AR and ARA_{70}. This precipitation was specific, as CW2 did not precipitate two other proteins (RXR and TR4 orphan receptor) incubated with AR. A Far-Western assay also demonstrated that ARA_{70} can bind to the immobilized ligand-binding domain of AR but not other control proteins. Together, these data suggest that an increase in the transcriptional activity by ARA_{70} is due to a direct interaction between AR and ARA_{70} (Yeh and Chang 1996).

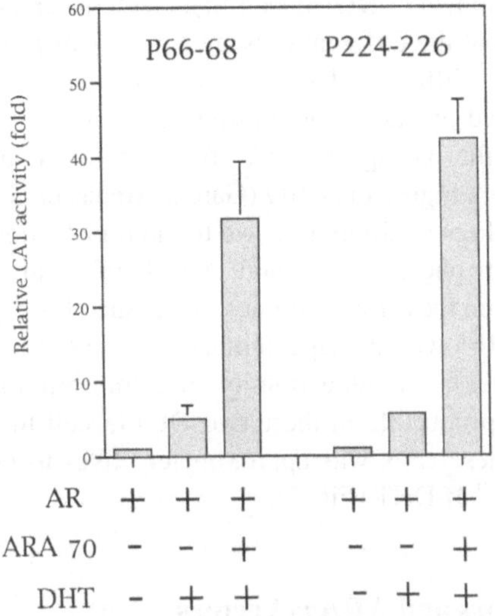

Fig. 1. ARA_{70} can enhance the transcriptional activity of the androgen receptor (*AR*) in different passage numbers of DU145 cells. Fixed amounts of pSG5-AR (1.5 μg) and pSG5-ARA70 (4.5 μg) were transfected into DU145 cells with passage numbers 66–68 and 224–226 in the presence of 10^{-9} *M* dihydrotestosterone (*DHT*) for a chloramphenicol acetyltransferase (*CAT*) assay. In each transfection, 3.5 μg of reporter [mouse mammary tumor virus chloramphenicol acetyltransferase (MMTV-ARE-CAT)] was co-transfected. A β-galactosidase expression plasmid, pCMV-β-gal, was used as an internal control for transfection efficiency. The total amount of DNA was adjusted to 10.5 μg with pSG5 in all experiments. The mock treatment was set as one-fold. All data were the average results ± S.D. of three independent experiments

2.2 DU145 Cell Lines with Distinct Morphology

Two DU145 cell lines (passage numbers 59–68 from American Type Culture Collection and passage numbers 205–226 from Dr. G. Wilding, University of Wisconsin-Madison) were used for our current ARA_{70} studies. Interestingly, we found differences in the morphology of these two DU145 cell lines. Using a Northern blot analysis of these two DU145 cell lines, we detected the expression of ARA_{70} mRNA contrary to our previous report (Yeh and Chang 1996) that suggested ARA_{70} mRNA levels were undetectable. The difference could be due to different passage numbers of DU145 cell lines with different morphology. It was reported that the different passage numbers of prostate cancer cell lines may have different characteristics. For example, Northern blot analysis detected prostatic acid phosphatase in prostate cancer LNCaP cell line with the passage numbers from 25 to 44, but not with the passage numbers higher than 100 (Garcia-Arenas et al. 1995). In addition to morphological differences, we found transfection efficiency (using the calcium phosphate method; Mizokami et al. 1994) was also different between these two cell lines, with much lower β-galactosidase activity in DU145 with passage numbers from 59 to 68. In spite of the above differences, the induction of AR transcriptional activity by ARA_{70} was reproducible in these two DU145 cell lines. The average induction in these cells was approximately five- to eight-fold in the presence of 10^{-9} *M* DHT (Fig. 1).

2.3 AR Vectors and ARA_{70} Vectors

To rule out the possibility that the effect of ARA_{70} resulted from the plasmid backbone sequence, the pSG5-AR expression vector was replaced with the pCMV-AR or pCMVneo-AR expression vector. As shown in Fig. 2a, ARA_{70} can induce the transcriptional activity of AR with different expression vectors in DU145 cells. Also, AR transcriptional activity was induced both with pSG5-ARA_{70} and pCMV-ARA_{70} (Fig. 2b). For a control, a parent expression vector was always used to obtain equal amounts of plasmids in each transfection. Together, these data indicate that DHT-dependent induction of the transcriptional activity of AR by ARA_{70} is not due to the backbone of the vectors.

2.4 Different ARE Reporters

We replaced MMTV-ARE CAT with PSA-ARE CAT, which encompasses the promoter of the prostate-specific antigen (PSA) gene from –540 bp to the translational start site. PSA is a classic androgen target gene and is widely used as a tumor marker for patients with prostate cancer. As shown in Fig. 3, ARA_{70} enhanced AR transcriptional activity on the PSA receptor gene five-fold in the presence of 10^{-9} *M* DHT. We

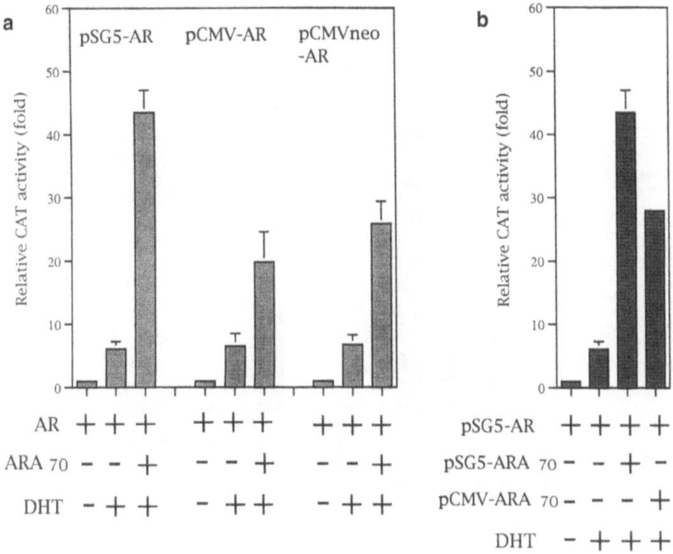

Fig. 2. Effects of ARA_{70} with different backbone plasmids of the androgen receptor (*AR*) (**a**) and *ARA_{70}* (**b**). Chloramphenicol acetyltransferase (*CAT*) activity was determined in DU145 cells co-transfected with (**a**) *pSG5-AR* (1.5 µg), *pCMV-AR* (1.5 µg), or *pCMVneo-AR* (1.5 µg), and *pSG5-ARA_{70}* (4.5 µg), or (**b**) *pSG5-AR* (1.5 µg) and *pSG5-ARA_{70}* (4.5 µg), or *pCMV-ARA_{70}* (4.5 µg) in the absence or presence of 10^{-9} *M* dihydrotestosterone (*DHT*). Mouse mammary tumor virus chloramphenicol acetyltransferase was used as a reporter for the assay of transcriptional activity. All data were the average results ± S.D. of three independent experiments

further applied two copies of ARE oligomers (tyrosine aminotransferase; TAT-ARE) linked to the CAT reporter to test ARA_{70} effect. While the CAT conversion rate was low, ARA_{70} still showed an induction effect (Fig. 3). These results suggest that ARA_{70} may be able to induce general AR target genes which may play an important role in the development of normal prostate and prostate cancer.

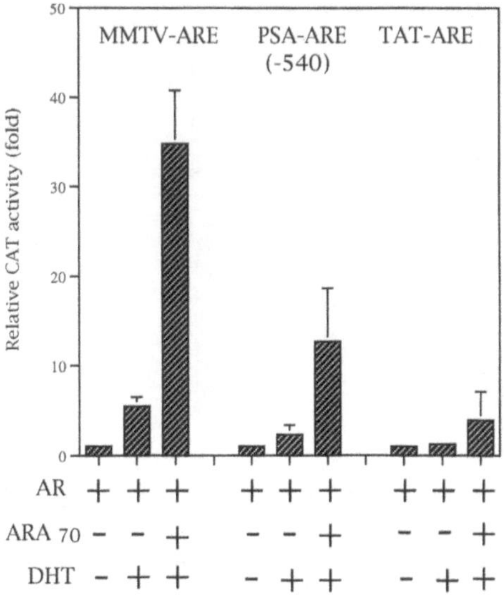

Fig. 3. Effects of ARA_{70} with different ARE reporters. Chloramphenicol acethyltransferase (*CAT*) activity was determined in DU145 cells co-transfected with pSG5-AR (1.5 μg), pSG5-ARA_{70} (4.5 μg), and 3.5 μg of mouse mammary tumor virus chloramphenicol acetyltransferase (*MMTV-ARE*) CAT, prostate-specific antigen ARE (*PSA-ARE*) CAT, or tyrosine aminotransferase ARE (*TAT-ARE*) CAT in the absence or presence of 10^{-9} *M* dihydrotestosterone (*DHT*). All data were the average results ± S.D. of three independent experiments

2.5 Mutated ARs

The effect of ARA_{70} on the transcriptional activity of mutant AR was also examined. It has been reported that some human prostate tumors were found to have mutations of the AR gene, and that a number of mutations in the hormone-binding domain can alter the specificity of AR (Taplin et al. 1995). Therefore, it was of great interest to examine whether ARA_{70} interacted with mutant ARs to understand any association between ARA_{70} and prostate cancer. In this study, three mutant ARs from human prostate tumors (AR715, AR874, and AR877) were tested. As shown in Fig. 4, the induction of transcriptional activity of these mutated ARs by ARA_{70} was similar to that of wild-type AR in the

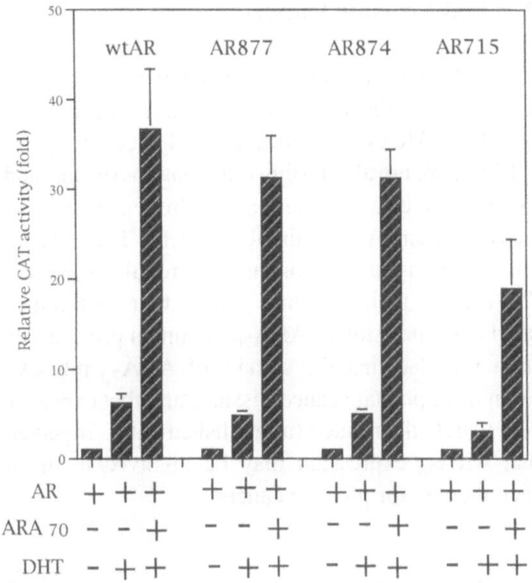

Fig. 4. Effects of ARA_{70} on the mutated androgen receptor (*AR*) transcriptional activity. Chloramphenicol acetyltransferase (*CAT*) activity was determined in DU145 cells co-transfected with pSG5-wild-type-AR (*wtAR*) (1.5 μg), pSVL-AR877 (*AR877*) (1.5 μg), pSVL-AR874 (*AR874*) (1.5 μg), or pSVL-AR715 (*AR715*) (1.5 μg), and pSG5-ARA_{70} (*ARA_{70}*) (4.5 μg) in the absence or presence of 10^{-9} *M* dihydrotestosterone (*DHT*). All data were the average results ± S.D. of three independent experiments

presence of 10^{-9} *M* DHT in DU145 cells (between six- and seven-fold). Another AR mutation with one amino acid substitution at the second zinc finger of DBD (Arg614 to His614) that proved to be insensitve to androgen action was demonstrated to be insensitive to ARA_{70} induction (data not shown). These results suggest that functional AR and ARA_{70} may be needed for the maximal androgen action in DU145 cells.

2.6 ARA_{70} and Prostate Cancer

We have recently reported that higher concentrations of hydroxyflutamide (HF), one of the antiandrogens, may become an agonist to androgens and that ARA_{70} may be able to enhance this agonistic activity (Yeh et al. 1997). We are also testing other antiandrogens, and the results indicate that agonist effect of these antiandrogens may be enhanced in the presence of AR and ARA_{70} in DU145 cells. These data suggest that agonist activity of the antiandrogens may require some co-activators, such as ARA_{70}, for their maximal action. Our preliminary data also demonstrated the expression of ARA_{70} in human prostate cancer. *In situ* hybridization revealed that the signals of ARA_{70} mRNA expression were found in most prostate cancer tissues, and that expression levels of ARA_{70} were widely distributed (unpublished data). Together, these data suggest that ARA_{70} expression may be involved in the response to antiandrogen therapy for prostate cancer.

2.7 Conclusion

Using several different controls, our data clearly demonstrate that ARA_{70} can increase the transcriptional activity of AR in DU145 cells by between three- and eight-fold. Since the induction can be detected only when AR and ARA_{70} in a ratio above 1:3 were co-transfected, it may occur in a very tightly controlled fashion. The correct conditions of transfection and ratios of AR and ARA_{70} may play a critical role in the induction of AR transcriptional activity. Further studies with antisense ARA_{70} or so-called dominant negative experiments, as well as knock-out ARA_{70} gene may prove that these *in vitro* effects of ARA_{70} can also occur *in vivo*. Until then, *in vitro* transfection data may only suggest a potential role of ARA_{70} in androgen-AR function.

In conclusion, ARA_{70} can induce the transcriptional activity of both wild-type and several mutant ARs in DU145 cells in a ligand-dependent manner. The induction occurs irrespective of plasmid backbone of AR or ARA_{70} vector, or reporter of the androgen target gene. Our data suggest that ARA_{70} may function as a relatively specific co-activator for AR. Although several co-factors have been demonstrated to interact with steroid receptors (review Horwitz et al. 1996), none of these co-factors have been reported to specifically enhance AR-mediated transcriptional activity. Therefore, it is likely that ARA_{70} has a different mechanism for interacting with AR. Further studies of the potential role of ARA_{70} are required to better understand the molecular mechanism of androgen action.

Acknowledgments. The plasmids of ARs with mutations in codons 715, 874, and 877 were kindly provided by Dr. Steven Balk.
This work was supported by NIH grants CA71570 and CA68568.
Authors H. Miyamoto and S. Yeh contributed equally to this work and should both be considered as first author of this chapter.

References

Chang C, Kokontis J, Liao ST (1988) Molecular cloning of human and rat complementary DNA encoding androgen receptors. Science 240:324–326

Evans RM (1988) The steroid and thyroid hormone receptor superfamily. Science 240:889–895

Garcia-Arenas R, Lin F-F, Lin D, Jin L-P, Shih CC-Y, Chang C, Lin M-F (1995) The expression of prostatic acid phosphate is transcriptionally regulated in human prostate carcinoma cells. Mol Cell Endocrinol 111:29–37

Horwitz KB, Jackson TA, Bain DL, Richer JK, Takimoto GS, Tung L (1996) Nuclear receptor coactivators and corepressors. Mol Endocrinol 10:1167–1177

Lubahn DB, Joseph DR, Sullivan PM, Willard HF, French FS, Wilson EM (1988) Cloning of human androgen receptor complementary DNA and localization to the X chromosome. Science 240:327–330

Mizokami A, Yeh S, Chang C (1994) Identification of 3',5'-cyclic adenosine monophosphate response element and other *cis*-acting elements in the human androgen receptor gene promoter. Mol Endocrinol 8:77–88

Taplin M-E, Bubley GJ, Shuster TD, Frantz ME, Spooner AM, Ogata GK, Keer HN, Balk SP (1995) Mutation of the androgen-receptor gene in metastatic androgen-independent prostate cancer. N Engl J Med 332:1393–1398

Yeh S, Chang C (1996) Cloning and characterization of a specific coactivator, ARA_{70}, for the androgen receptor in human prostate cells. Proc Natl Acad Sci USA 93:5517–5521

Yeh S, Miyamoto H, Chang C (1997) Hydroxyflutamide may not always be a pure antiandrogen. Lancet 349:852–853

3 Role of Coactivators in Transcriptional Activation by Estrogen Receptors

M.G. Parker, D. Heery, E. Kalkhoven, and J. Valentine

3.1 Introduction

Estrogens regulate the transcription of target genes either directly by binding to specific response elements, referred to as transactivation, or indirectly by modulating the activity of other transcription factors. These latter effects can be inhibitory (Stein and Yang 1995), referred to as transrepression, or stimulatory, such as in the case of AP-1 activation (Philips et al. 1993; Webb et al. 1995). Two distinct transcriptional activation functions have been identified: AF-1 in the N-terminal domain, whose activity may be modulated by phosphorylation (Ali et al. 1993; Bunone et al. 1996; Kato et al. 1995), and AF-2 in the hormone-binding domain, which is induced by hormone binding. A short amphipathic α-helix in the C-terminal part of the ligand-binding domain, conserved in all transcriptionally active nuclear receptors, is essential for AF-2 function (Danielian et al. 1992). This helix, which corresponds to helix 12 in the ligand-binding domain of nuclear receptors (Bourguet et al. 1995; Renaud et al. 1995; Wagner et al. 1995; Wurtz et al. 1996), is aligned across the core of the ligand-binding domain in the estrogen receptor (Brzozowski et al. 1997). In conjunction with conserved residues in helix 3 (Henttu et al. 1997), it appears to form part of a surface which interacts with downstream target proteins required for AF-2 activity. Some of these receptor interacting proteins (RIPs) function as coactivators, but the roles of many others are unclear (for a review see Heery and Parker 1997). This short review will summarise what is known about the RIP160 family of coactivators.

3.2 Identification of Target Proteins Recruited by Activated Receptors

Estrogen receptors have been shown to bind directly to basal transcription factors in vitro but the significance of these interactions is unclear since they are unaffected by mutations in the receptor that destroy transcriptional activity (Jacq et al. 1994; Sadovsky et al. 1995). The involvement of additional targets is suggested by the observation that AF-2 activity is inhibited when receptors are overexpressed, suggesting that limiting downstream target proteins required for gene transcription are being sequestered (Tasset et al. 1990). A number of candidate target proteins that bind to the ligand-binding domain of receptors in a ligand-dependent manner (Cavaillès et al. 1994; Halachmi et al. 1994) have been detected in mammalian cell extracts. Since their interaction with mutant receptors correlates with the transcriptional activity of the mutants they may have a role in ligand-dependent transcriptional activation by AF-2.

Receptor-interacting proteins may be classified into two groups: The first group consists of two families of proteins which appear to stimulate transcription and represent bona fide coactivators (Table 1). One family, originally identified as proteins of 160 kDa, consists of a number of related proteins; the steroid receptor coactivator proteins (SRC-1a and SRC-1e; Kamei et al. 1996; Oñate et al. 1995), transcription intermediary factor (TIF2; Voegel et al. 1996; independently identified as GRIP-1; Hong et al. 1996) and CBP interacting protein (p/CIP; Torchia et al. 1997), also named ACTR (Chen et al. 1997), AIB1 (Anzick et al. 1997), and RAC3 (Li et al. 1997). The binding of these proteins to nuclear receptors in vitro is reduced or abolished when mutations are introduced

Table 1. Coactivators for activated nuclear receptors

Coactivator	Mol wt	Receptor binding	Reference
SRC-1/NCoA1	157	PR, ER, TR, RXR, GR	Oñate et al. (1995) Kamei et al. (1996)
TIF2/GRIP1	158	RAR, RXR, ER, TR	Voegel et al. (1996)
pCIP/AIB1/ACTR	155	RAR, RXR, ER, TR	Torchia et al. (1996) Anzick et al. (1997) Chen et al. (1997)
CBP/p300	260	RAR, RXR, ER, TR	Kamei et al. (1996) Chakravarti et al. (1996) Yao et al. (1996) Hanstein et al. (1996)

PR, progesterone receptor; *ER,* estrogen receptor; *TR,* thyroid hormone receptor; *RAR,* retinoic acid receptor; *RXR,* retinoid X receptor; *GR,* glucocorticoid receptor.

into the conserved hydrophobic residues in helix 12 (Kamei et al. 1996) or the conserved lysine in helix 3 (Henttu et al. 1997) which impair AF-2 activity. This suggests that, in the presence of ligand, helices 3 and 12 form a composite surface recognised by the p160 proteins. The second family comprises CBP and p300, which were originally shown to function as coactivators for CREB, the transcription factor that mediates responses to protein kinase A stimulation. Subsequently, however, CBP/p300 were shown to function as coactivators for many other transcription factors and may play a central role in many signalling pathways (Janknecht and Hunter 1996; Shikama et al. 1997). The p160 and CBP/p300 families of proteins are considered bona fide coactivators, firstly because they potentiate the transcriptional activity of nuclear receptors in transfected cells, and secondly because dominant negative versions of the protein or specific microinjected antibodies specific for the coactivators block nuclear receptor signalling (Oñate et al. 1995; Torchia et al. 1997). The recruitment of these proteins by activated receptors is not straightforward, however, because, in addition to their direct interaction with receptors, the RIP160 and CBP/p300 families also interact with one another (Hanstein et al. 1996; Kamei et al. 1996; Yao et al. 1996). Thus it is not clear whether members of the two families are recruited independently or sequentially to activated receptors and whether both contact the helix 3/12 surface simultaneously.

Another group of unrelated proteins have been shown to interact with the ligand-binding domain of receptors but which do not significantly stimulate receptor mediated transcription in a co-transfection assay and therefore their role in receptor function is still unclear. This group includes a 140-kD protein termed RIP 140 (Cavaillès et al. 1995), TIF1 (Le Douarin et al. 1995) and TRIP1/SUG1 (Lee et al. 1995; vom Baur et al. 1996). RIP140 stimulates receptor-mediated transcription in yeast (Joyeux et al. 1996) and contains an "activation domain" which is functional in mammalian cells (L'Horset et al. 1996), but it is still unclear whether it is a bona fide coactivator. TIF1 has been shown to interact with proteins associated with heterochromatin and may play a role in chromatin remodelling (Le Douarin et al. 1996).

3.3 Structure and Function of Coactivators for Nuclear Receptors

To date, it appears that the p160 proteins and CBP/p300 are the most important of the receptor-interacting proteins in terms of their ability to potentiate the transcriptional activity of receptors. The p160 proteins, namely SRC-1, TIF2/GRIP1 and pCIP/ACTR/AIB1/RAC3 share four regions of homology, corresponding to functionally conserved domains, and conserved Ser/Thr and glutamine-rich regions (Fig. 1). At the N-terminus is a region which is homologous to the PAS and helix-loop-helix domains first noted in the transcription factors Per, Arnt and Sim. In these proteins the domains appear to play a role in DNA binding and dimerisation but their function in the coactivator family is unknown.

In the middle of the proteins is a nuclear receptor-binding domain containing three conserved LXXLL motifs (where L is leucine and X is any amino acid) which is both necessary and sufficient to mediate receptor binding (Heery et al. 1997). Multiple copies of the motif are also present in CBP/p300 and other receptor-interacting proteins (Heery et al. 1997; Torchia et al. 1997). The LXXLL motifs are predicted to be helical with as few as eight amino acids being sufficient for binding to activated receptors. The motifs are capable of binding multiple receptors but preliminary analysis suggests that they may be selectively

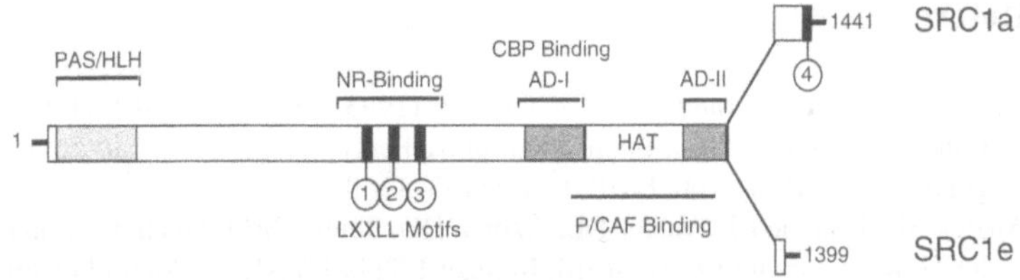

Fig. 1. Functional domains in steroid receptor coactivator 1a (*SRC-1a*) and *SRC-1e*. The *PAS/HLH* domain shares homology with a region of Per, Arnt and Sim implicated in dimerisation and DNA binding. The nuclear receptor (*NR*) binding region consists of three LXXLL motifs *1*, *2* and *3*. The activation domain, *AD-I*, colocalises with a CBP interaction domain. Histone acetyl transferase (*HAT*) and *P/CAF* binding activity have been reported by Spencer et al. (1997). A second activation domain, *AD-II*, is located at the C-terminus (Kalkhoven et al. 1997)

utilised by different receptors. Since peptides corresponding to the LXXLL motif inhibit the binding of SRC-1 by the activated estrogen receptor (Heery et al. 1997), this type of interaction may represent a target for therapeutic intervention.

Towards the C-terminus are two activation domains, AD-1 and AD-II. AD-I colocalises with a region that binds CBP and it is likely that its ability to stimulate transcriptional activity is mediated by CBP. In contrast, the target of AD-II is unknown. In addition, SRC-1 and ACTR encode histone acetyl transferases (HATs) and are capable of recruiting P/CAF, another histone acetyl transferase (Spencer et al. 1997; Yang et al. 1996). Thus it is clear that the p160 family of proteins are multidomain proteins that may serve several functions. Firstly, SRC-1 may play a role in chromatin remodelling by acetylating histones, but it is unclear why three different HATs, SRC-1, CBP and P/CAF, should be recruited to target promoters if they serve the same role. One possibility is that histones may not be the only target for the HATs in different families of coactivators, in which case they may serve distinct functions. Secondly, they may play a role in the recruitment of the basal transcription machinery since CBP has been shown to interact with TFIIB. It is conceivable that AD-II has a similar role but we have yet to identify its target.

References

Ali S, Metzger D, Bornert J-M, Chambon P (1993) Modulation of transcriptional activation by ligand-dependent phosphorylation of the human oestrogen receptor A/B region. EMBO J 12:1153–1160

Anzick SL, Kononen J, Walker RL, Azorsa DO, Tanner MM, Guan X-Y, Sauter G, Kallioniemi O-P, Trent JM, Meltzer PS (1997) AIB1, a steroid receptor coactivator amplified in breast and ovarian cancer. Science 277:965–968

Bourguet W, Ruff M, Chambon P, Gronemeyer H, Moras D (1995) Crystal-structure of the ligand-binding domain of the human nuclear receptor RXR-alpha. Nature 375:377–382

Brzozowski AM, Pike ACW, Dauter Z, Hubbard RE, Bonn T, Engstrom O, Ohman L, Greene GL, Gustafsson J-α, Carlquist M (1997) Molecular basis of agonism and antagonism in the oestrogen receptor. Nature 389:753–758

Bunone G, Briand P-A, Miksicek RJ, Picard D (1996) Activation of the unliganded estrogen receptor by EGF involves the MAP kinase pathway and direct phosphorylation. EMBO J 15:2174–2183

Cavaillès V, Dauvois S, Danielian PS, Parker MG (1994) Interaction of proteins with transcriptionally active estrogen receptors. Proc Natl Acad Sci USA 91:10009–10013

Cavaillès V, Dauvois S, L'Horset F, Lopez G, Hoare S, Kushner PJ, Parker MG (1995) Nuclear factor RIP140 modulates transcriptional activation by the estrogen receptor. EMBO J 14:3741–3751

Chen H, Lin RJ, Schiltz RL, Chakravarti D, Nash A, Nagy L, Privalsky ML, Nakatani Y, Evans RM (1997) Nuclear receptor coactivator ACTR is a novel histone acetyltransferase and forms a multimeric activation complex with P/CAF and CBP/p300. Cell 190:569–580

Danielian PS, White R, Lees JA, Parker MG (1992) Identification of a conserved region required for hormone dependent transcriptional activation by steroid hormone receptors. EMBO J 11:1025–1033

Halachmi S, Marden E, Martin G, MacKay H, Abbondanza C, Brown M (1994) Estrogen receptor-associated proteins – possible mediators of hormone-induced transcription. Science 264:1455–1458

Hanstein B, Eckner R, DiRenzo J, Halachmi S, Liu H, Searcy B, Kurokawa R, Brown M (1996) p300 is a component of an estrogen receptor coactivator complex. Proc Natl Acad Sci USA 93:11540–11545

Heery DM, Parker MG (1997) Ligand-induced transcription by nuclear receptors. Retinoids 13:26–30

Heery DM, Kalkhoven E, Hoare S, Parker MG (1997) A signature motif in transcriptional co-activators mediates binding to nuclear receptors. Nature 387:733–736

Henttu PMA, Kalkhoven E, Parker MG (1997) AF-2 activity and recruitment of steroid receptor coactivator 1 to the estrogen receptor depend on a lysine residue conserved in nuclear receptors. Mol Cell Biol 17:1832–1839

Hong H, Kohli K, Trivedi A, Johnson DL, Stallcup MR (1996) Grip1, a novel mouse protein that serves as a transcriptional coactivator in yeast for the hormone-binding domains of steroid-receptors. Proc Natl Acad Sci USA 93:4948–4952

Jacq X, Brou C, Lutz Y, Davidson I, Chambon P, Tora L (1994) Human $TAF_{II}30$ is present in a distinct TFIID complex and is required for transcriptional activation by the estrogen receptor. Cell 79:107–117

Janknecht R, Hunter T (1996) A growing coactivator network. Nature 383:22–23

Joyeux A, Cavaillès V, Balaguer P, Nicolas JC (1997) RIP 140 enhances nuclear receptor-dependent transcription in vivo, in yeast. Mol Endocrinol 11:193–202

Kalkhoven E, Valentine JE, Heery DM, Parker MG (1998) Isoforms of steroid receptor coactivator 1 differ in their ability to potentiate transcription by the oestrogen receptor. EMBO J 17:232–243

Kamei Y, Xu L, Heinzel T, Torchia J, Kurokawa R, Gloss B, Lin S-C, Heyman RA, Rose DW, Glass CK, Rosenfeld MG (1996) A CBP integrator complex mediates transcriptional activation and AP-1 inhibition by nuclear receptors. Cell 85:403–414

Kato S, Endoh H, Masuhiro Y, Kitamoto T, Uchiyama S, Sasaki H, Masushige S, Gotoh Y, Nishida E, Kawashima H, Metzger D, Chambon P (1995) Activation of the estrogen-receptor through phosphorylation by mitogen-activated protein kinase. Science 270:1491–1494

Le Douarin B, Zechel C, Garnier JM, Lutz Y, Tora L, Pierrat B, Heery D, Gronemeyer H, Chambon P, Losson R (1995) The N-terminal part of TIF1, a putative mediator of the ligand-dependent activation function (AF-2) of nuclear receptors, is fused to B-raf in the oncogenic protein T18. EMBO J 14:2020–2033

Le Douarin B, Nielsen AL, Garnier J-M, Ichinose H, Jeanmougin F, Losson R, Chambon P (1996) A possible involvement of TIF1a and TIF1b in the epigenetic control of transcription by nuclear receptors. EMBO J 15:6701–6715

L'Horset F, Dauvois S, Heery DM, Cavaillès V, Parker MG (1996) RIP-140 interacts with multiple nuclear receptors by means of two distinct sites. Mol Cell Biol 16:6029–6036

Lee JW, Ryan F, Swaffield JC, Johnston SA, Moore DD (1995) Interaction of thyroid-hormone receptor with a conserved transcriptional mediator. Nature 374:91–94

Li H, Gomes PJ, Chen JD (1997) RAC3, a steroid/nuclear receptor-associated coactivator that is related to SRC-1 and TIF2. Proc Natl Acad Sci USA 94:8479–8484

Oñate SA, Tsai SY, Tsai M-J, O'Malley BW (1995) Sequence and characterization of a coactivator for the steroid hormone receptor superfamily. Science 270:1354–1357

Philips A, Chalbos D, Rochefort H (1993) Estradiol increases and anti-estrogens antagonize the growth factor-induced activator protein –1 activity in MCF7 breast cancer cells without affecting c-fos and c-jun synthesis. J Biol Chem 268:14103–14108

Renaud J-P, Rochel N, Ruff M, Vivat V, Chambon P, Gronemeyer H, Moras D (1995) Crystal structure of the RAR-γ ligand-binding domain bound to all-trans retinoic acid. Nature 378:681–689

Sadovsky Y, Webb P, Lopez G, Baxter JD, Cavaillès V, Parker MG, Kushner PJ (1995) Transcriptional activators differ in their responses to overexpression of TATA-box-binding protein. Mol Cell Biol 15:1554–1563

Shikama N, Lyon J, La Thangue NB (1997) The p300/CBP family: integrating signals with transcription factors and chromatin. Trends Cell Biol 7:230–236

Spencer TE, Jenster G, Burcin MM, Allis CD, Zhou J, Mizzen CA, McKenna NJ, Oñate SA, Tsai SY, Tsai M-J, O'Malley BW (1997) Steroid receptor coactivator-1 is a histone acetyltransferase. Nature 389:194–198

Stein B, Yang MX (1995) Repression of the interleukin-6 promoter by estrogen receptor is mediated by NF-κ and C/EBPβ. Mol Cell Biol 15:4971–4979

Tasset D, Tora L, Fromental C, Scheer E, Chambon P (1990) Distinct classes of transcriptional activating domains function by different mechanisms. Cell 62:1177–1187

Torchia J, Rose DW, Inostroza J, Kamei Y, Westin S, Glass CK, Rosenfeld MG (1997) The transcriptional co-activator p/CIP binds CBP and mediates nuclear-receptor function. Nature 387:677–684

Voegel JJ, Heine MJS, Zechel C, Chambon P, Gronemeyer H (1996) TIF2, a 160-kDa transcriptional mediator for the ligand-dependent activation function AF-2 of nuclear receptors. EMBO J 15:101–108

vom Baur E, Zechel C, Heery D, Heine MJS, Garnier JM, Vivat V, Le Douarin B, Gronemeyer H, Chambon P, Losson R (1996) Differential ligand-dependent interactions between the AF-2 activating domain of nuclear receptors and the putative transcriptional intermediary factors mSUG1 and TIF1. EMBO J 15:110–124

Wagner RL, Apriletti JW, McGrath ME, West BL, Baxter JD, Fletterick RJ (1995) A structural role for hormone in the thyroid hormone receptor. Nature 378:690–697

Webb P, Lopez GN, Uht RM, Kushner PJ (1995) Tamoxifen activation of the estrogen receptor/AP-1 pathway: potential origin for the cell specific estrogen-like effects of antiestrogens. Mol Endocrinol 9:443–456

Wurtz J-M, Bourguet W, Renaud J-P, Vivat V, Chambon P, Moras D, Gronemeyer H (1996) A canonical structure for the ligand-binding domain of nuclear receptors. Nat Struct Biol 3:87–94

Yang X-J, Ogryzko VV, Nishikawa J-I, Howard BH, Nakatani Y (1996) A p300/CBP-associated factor that competes with the adenoviral oncoprotein E1 A. Nature 382:319–324

Yao T-P, Ku G, Zhou N, Scully R, Livingston DM (1996) The nuclear hormone receptor coactivator SRC-1 is a specific target of p300. Proc Natl Acad Sci USA 93:10626–10631

4 A Search for a Mechanism for the Antiinflammatory Action of Glucocorticoids

A.C.B. Cato

4.1 Introduction

Cortisol, a glucocorticoid hormone, was first isolated from adrenal tissues in 1936 (Mason et al. 1936a,b) and about 10 years later it was shown to be effective in the treatment of rheumatoid arthritis, particularly when administered pharmacologically (Hench et al. 1949, 1950).

Since then glucocoticoids on the whole have been recognized as highly efficient for the treatment of various inflammatory diseases. Their long-term use, however, is accompanied by adverse side effects ranging from emotional lability and cushingoid habitus to osteoporosis

(Lukert and Raisz 1990; Piper et al. 1991). It is therefore important to understand their mode of action in order to try and separate their beneficial from their unwanted effects.

Glucocorticoids belong to the family of steroid hormones that enter the cell by passive diffusion and bind to receptor molecules, converting them from an inactive complex into an efficient transcription factor. Like other steroid receptors, upon ligand binding, the glucocorticoid receptor (GR) acquires the ability to either positively or negatively regulate the expression of genes. While a lot is known about how the GR transactivates, it is only recently that details of transrepression are being disclosed. Nevertheless, negative regulation is considered important for the antiinflammatory action of glucocorticoids since most cytokine and chemokine genes activated during chronic inflammation are negatively regulated by the GR (Cato and Wade 1996). As these proinflammatory genes are mostly under the control of the transcription factors activator protein 1 (AP-1), nuclear factor-κB (NF-κB), CCAAT/enhancer binding protein β (C/EBPβ), and nuclear factor of activated T-cells (NF-AT), negative regulation of activity of these transcription factors by the GR has become a paradigm for the antiinflammatory action of glucocorticoids.

However, negative effects of glucocorticoids on the synthesis of proinflammatory cytokines and chemokines are not always at the transcriptional level. Several reports have demonstrated that post-transcriptional mechanisms are equally involved in the glucocorticoid-mediated suppression of activity of these proteins. These post-transcriptional regulatory processes can be classified as glucocorticoid-mediated destabilization of cytokine mRNA, glucocorticoid-induced protein degradation, and glucocorticoid-mediated interference of signal transduction mechanisms by proinflammatory cytokines and chemokines.

In this chapter I will discuss various regulatory activities of the GR with a view to determining which of them are most relevant to the antiinflammatory action of glucocorticoids.

4.2 Transcriptional Repression

Glucocorticoids may exert their antiinflammatory action by inhibiting the activity of a few transcription factors. Table 1 lists a number of genes involved in inflammatory processes that are transcriptionally repressed by the GR. It is clear from this list that only a few transcription factors (AP-1, NF-κB, NF-AT, and c/EBPβ) are involved in this negative action of the GR.

Several lines of evidence suggest that downregulation of the activity of these transcription factors inhibit inflammation. For example, compounds isolated as inhibitors of NF-κB activity exhibited potent antiinflammatory activity in different animal models for inflammation (Pierce et al. 1997). Furthermore, local administration of antisense phosphorothioate oligonucleotides targeted against the translational start site of the p65 subunit of NF-κB abrogated colitis induced in mice by treatment with 2, 4, 6-trinitrobenzene sulfonic acid (Neurath et al. 1996). These results demonstrate that downregulation of activity of NF-κB contributes significantly to antiinflammatory action. The inhibition of NF-κB activity by the GR is therefore one of the bases for the antiinflammatory action of glucocorticoids.

How does the GR negatively regulate the activity of the various transcription factors? The exact mechanism is still unknown but it has been shown that DNA binding activity of both NF-κB and AP-1 is inhibited by the GR in in vitro receptor-DNA binding studies (Diamond et al. 1990; Yang-Yen et al. 1990; Ray and Prefontain 1994; Scheinman et al. 1995). However, in the case of AP-1, in vivo footprinting studies showed no change in the occupancy of the AP-1 binding site in the presence of the activated GR. This indicates that, for negative regulation, the GR does not displace AP-1 from its binding site, but rather it is tethered onto the DNA-bound form of this transcription (König et al. 1992). In vivo footprinting of NF-κB in the presence of the GR has not yet been done but if it turns out to produce similar results to those with AP-1, this would mean that the GR uses similar mechanisms in its negative regulation of the action of these two transcription factors.

In another model for GR-mediated inhibition of activity of NF-κB, the receptor was reported to interfere with translocation of the p65 but not the p50 subunit of NF-κB from cytoplasm into nucleus following cytokine action (Ohtsuka et al. 1996). These results were obtained in immunoblot assays from analyses of nuclear and cytoplasmic proteins. There is, therefore, the need for confirmation by careful immunofluorescence experiments.

Table 1. A list of proinflammatory genes that are negatively regulated at the transcriptional level by the glucocorticoid receptor (GR)

Genes	Regulatory factors	Effect of GR	References
Adhesion molecules			
ICAM 1	NF-κB	Negative	Cronstein et al. (1992); Van de Solpe et al. (1994); Caldenhoven et al. (1995)
E-Selectin	NF-κB, NF-ELAM-1, NF-ELAM-2, HMGI (Y), ATF-2	Negative	Cronstein et al. (1992); De Lucas et al. (1994); Lewis et al. (1994); Schindler and Baichwal (1994); Whitley et al. (1994)
VCAM	NF-κB	Unknown	Landemarco et al. (1992)
LFA-1	Unknown	Negative	Pitzalis et al. (1997)
Cytokines/chemokines			
IL-2	NF-ATp, AP-1	Negative	Vacca et al. (1992); Paliogianni et al. (1993b)
IL-4	NF-AT	Unknown	Chuvpilo et al. (1993); Szabo et al. (1993)
IL-5	NF-κB, AP-1	Negative	Mori et al. (1997)
IL-6	NF-κB, c/EBPβ	Negative	Ray et al. (1991); Stein and Yang (1995)
IL-8	NF-κB	Negative	Mukaida et al. (1994); Okamoto et al. (1994)
IL-1β	c/EBPβ, CREB	Negative	Lee et al. (1988)
RANTES	NFκB, c/EBPβ	Negative	Kwon et al. (1995); Nelson et al. (1996); Ortiz et al. (1996); Wingett et al. (1996)
Interferon-β	NF-κB, HMGI(Y)	Unknown	Thanos and Maniatis (1992)
Enzymes			
iNOS	NF-κB	Negative	Radomski et al. (1990); Xie et al. (1994); Kleinert et al. (1996)
Cox-2	NF-κB, c/EBPβ	Negative	O'Banion et al. (1992); Mitchell et al. (1994); Yamamoto et al. (1995)

Table 1 *(continued)*

Genes	Regulatory factors	Effect of GR	References
Collagenase I	AP-1	Negative	Jonat et al. (1990, 1992)
Collagenase IV (92 kDa)	AP-1	Negative	Huhtala et al. (1991); Oikarinen et al. (1993)
Other factors			
MHC class II gene (1Aβ)	Unknown	Negative	Celada et al. (1993)

ICAM, intercellular adhesion molecule; NF, nuclear factor; ELAM, endothelial/leukocyte adhesion molecule; iNOS, inducible nitric-oxide synthase; HMG, high mobility group protein; CREB, cAMP response element binding protein; AP, activator protein; c/EBP, CCAAT/enhancer-binding protein β; ATF, activating transcription factor; Cox-2, cyclooxygenase 2; LFA-1, lymphocyte function-associated antigen 1; V-CAM, vascular adhesion cell molecule; RANTES, regulated on activation, normal T-cell expressed and secreted.

Recently, more than 200 synthetic glucocorticoids have been screened based on their ability to repress the activity of AP-1 and NF-κB, and to inhibit inflammation in animal models. Three substances have been identified that mediate transrepression but are compromised in their ability to transactivate. Interestingly, these three glucocorticoid analogues proved quite effective as antiinflammatory drugs (Vayssière et al. 1997). Conversely, in an independent study, glucocorticoid analogues that were defective in their ability to repress NF-κB-regulated genes turned out to have reduced antiinflammatory activity (Heck et al. 1997). These findings together confirm that the ability of the GR to repress the activity of transcriptional factors such as AP-1 and NF-κB contributes to the antiinflammatory action of glucocorticoids.

4.3 Post-transcriptional Repression

In addition to the repression of transcription, the GR negatively regulates the synthesis of several proinflammatory cytokines and chemokines through post-transcriptional mechanisms. Table 2 lists examples of

Table 2. A list of proinflammatory cytokines and chemokines that are negatively regulated by the glucocorticoid receptor (GR) via destabilization of their mRNA

Genes	Effect of GR on mRNA stability	References
Cytokines/chemokines		
GM-CSF	Destabilization	Tobler et al. (1992)
IL-8	Destabilization	Tobler et al. (1992)
IL-6	Destabilization	Tobler et al. (1992); Amano et al. (1993)
IL-2	Destabilization	Boumpas et al. (1991)
IL-1β	Destabilization	Lee et al. (1988); Amano et al. (1993)
Enzymes		
iNOS	Destabilization	Kunz et al. (1996)

negative regulation of cytokine and chemokine gene expression through glucocorticoid-mediated mRNA destabilization.

Most studies aimed at determining the mechanism by which GR inhibits interleukin (IL)-2 production by antigen-stimulated T cells have focused on inhibition of gene transcription. However, other mechanisms are also involved. One of them is glucocorticoid-mediated downregulation of IL-2 expression by decreasing the stability of IL-2 mRNA (Boumpas et al. 1991).

Glucocorticoids also downregulate the expression of the genes coding for granulocyte-macrophage colony-stimulating factor (GM-CSF), IL-8, and IL-6 mainly by decreasing the stability of their mRNA (Tobler et al. 1992). The molecular mechanisms involved in this destabilization of mRNA are not known. However, cytokines prone to rapid up- and downregulation such as GM-CSF, IL-6, and IL-8 possess a repeated AUUUA sequence in their 3′-untranslated region. This sequence is thought to bind proteins that may promote degradation of the mRNA. Whether the GR controls the synthesis of these proteins is at the moment unknown.

In addition to mRNA destabilization, glucocorticoids have other negative effects on several genes that are involved in inflammation. Endotoxin-induced tumor necrosis factor (TNF)-α expression in monocytes is inhibited by glucocorticoids at the level of translation of TNFα mRNA, while the transcriptional rate is only marginally decreased (Beutler et al. 1986; Han et al. 1990).

Glucocorticoids also have pronounced inhibitory effects on the inducible isoform of nitric oxide synthase (iNOS), an enzyme that plays an important role in vascular homeostasis and inflammation (Kunz et al. 1996; Walker et al. 1997). In cardiac microvascular endothelium, glucocorticoids downregulate iNOS by limiting the availability of iNOS co-factors or substrates needed in the L-arginine–nitric oxide pathway (for reviews see Nathan 1992; Moncada and Higgs 1993). This occurs through the suppression of cytokine-induced increases in guanosine triphosphate (GTP) cyclohydrolase I expression and intracellular tetrahydrobiopterin content (Simmons et al. 1996). Glucocorticoids also block the increase in L-arginine transport by suppressing the induction of high affinity cationic amino acid transport 1 and 2B and the low affinity 2 A transporter (Simmons et al. 1996). In addition, they inhibit cytokine induction of argininosuccinate synthase (Simmons et al. 1996), the rate-limiting enzyme for the de novo synthesis of arginine from citrulline. Thus in cardiac microvascular endothelial cells, downregulation of iNOS at the transcription level by glucocorticoids is only marginal compared with the many effects of this hormone on the availability of substrate and co-factors for iNOS activity.

4.4 Negative Regulation of Signal Transduction Pathways

The GR can negatively regulate the action of protein tyrosine kinases that are necessary for the action of a number of proinflammatory cytokines and chemokines. The earliest consequence of T-cell receptor (TCR) cross-linking in TCR signaling is the activation of tyrosine kinases including Lck and ZAP-70. ZAP-70 interacts with Vav, a SH2 domain-containing protein that is rapidly phosphorylated at tyrosine residues upon T-cell activation. Vav in turn interacts through its SH2 domain with SLP-76, a tyrosine-phosphorylated protein that also binds the adaptor protein Grb2. This protein leads to the activation of the Ras-Raf-mitogen-activated protein (MAP) kinase pathway and the activation of transcription factors that control the expression of c-Fos which, together with c-Jun, constitute AP-1, a positive regulator of IL-2 promoter activity (for reviews see Rao et al. 1997; Koretzky 1997) (Fig. 1).

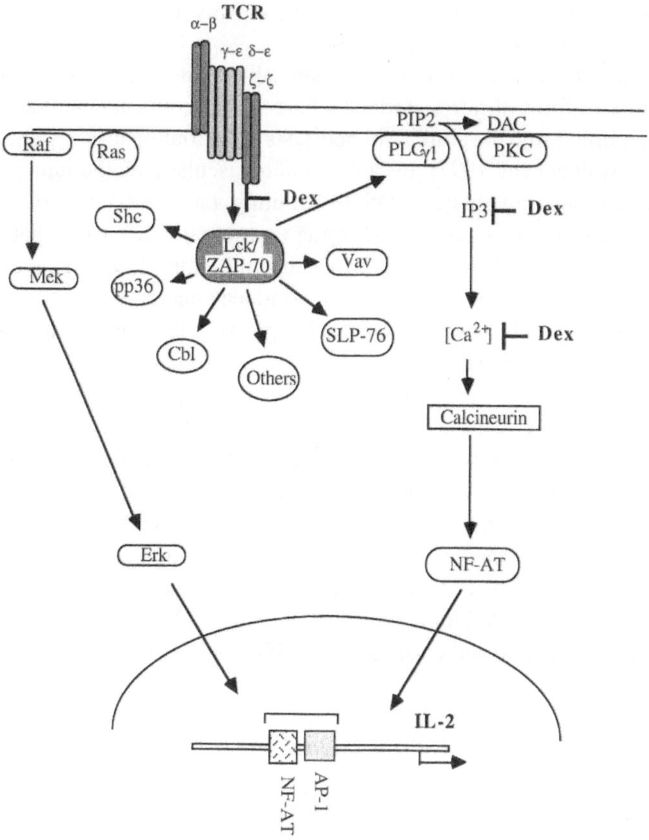

Fig. 1. A simplified view of some of the signaling pathways emanating from T-cell receptor (*TCR*) cross-linking. The steps reported to be inhibited by the glucocorticoid receptor have been indicated with *Dex* (the synthetic glucocorticoid dexamethasone). *PLCγ1*, phospholipase-Cγ1; *PKC*, protein kinase C; *IL-2*, interleukin 2; *NF-AT*, nuclear factor of activated T-cells; *AP-1*, activator protein 1

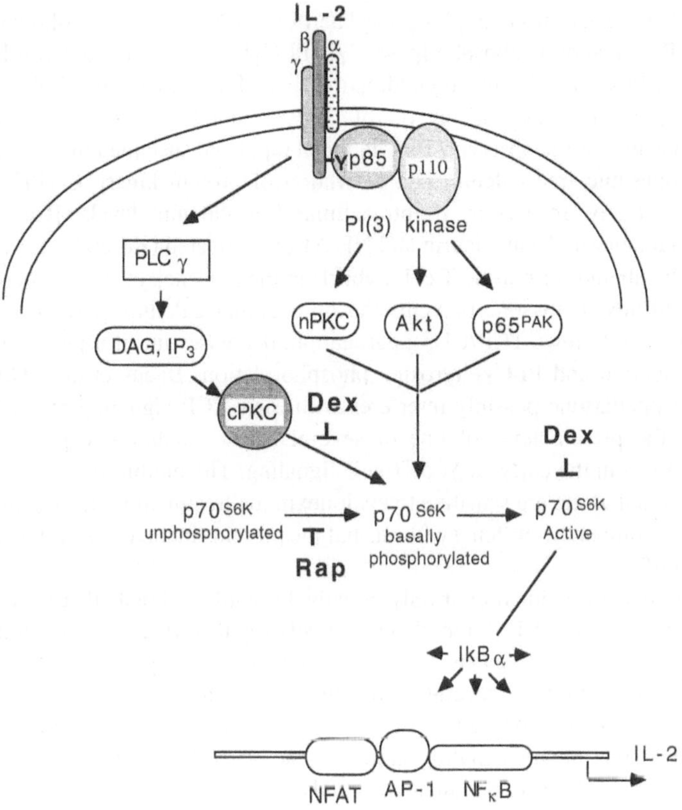

Fig. 2. Interleukin-2 (*IL-2*) receptor-coupled signaling pathways. The steps inhibited by glucocorticoid and by the immunosuppressant rapamycin have been indicated as *Dex* (the synthetic glucocorticoid dexamethasone) and *Rap*, respectively. *PLCγ*, phospholipase-Cγ; *PKC*, protein kinase C; *IkBα*, inhibitor protein α of NFκB; *NF-AT*, nuclear factor of activated T-cells; *AP-1*, activator protein 1; *NFκB*, nuclear factor κ B

One of the proteins phosphorylated on tyrosine residues following TCR activation is phospholipase-Cγ1 (PLCγ1). This results in hydrolysis of PI 4,5-bisphosphate yielding the second messengers inositol 1, 4, 5-triphosphate and diacylglycerol. These second messengers are responsible for the observed TCR-induced rapid and sustained increase in cytoplasmic free calcium and activation of protein kinase C (PKC), respectively. In T-cells, if intracellular free calcium levels, $[Ca^{2+}]_i$, remain elevated, calcineurin and NF-AT are activated (Koretzky 1997).

Incubation of murine T-cell hybrids in the presence of the glucocorticoid dexamethasone prevents the intracellular calcium increase that normally follows TCR/CD3 aggregation, decreases inositol phosphate production and PLCγ1 tyrosine phosphorylation (Bacus et al. 1996). Dexamethasone possibly interferes with early TCR signaling by altering the protein level of one or several kinases and/or phosphatases involved in the early steps of T-cell signaling. The inhibitory effects of dexamethasone are sensitive to cycloheximide treatment indicating that they require new protein synthesis but the proteins involved are yet to be identified.

Glucocorticoids do not only negatively regulate signals that lead to the production of IL-2 but also negatively regulate IL-2 receptor function. This is achieved by downregulation of IL-2-dependent tyrosine phosphorylation of several intracellular proteins (Paliogianni et al. 1993a). In addition, glucocorticoids inhibit the phosphorylation of $p70^{s6k}$, one of the signaling molecules activated by the IL-2 receptor through PI(3) kinase. Glucocorticoids further antagonize cPKC-induced activation of $p70^{s6k}$ (Monfar and Blenis 1996) (Fig. 2).

So far the way dexamethasone represses $p70^{s6k}$ phosphorylation is not known. It is not a result of a primary response of the hormone since it is repressed by inhibitors of transcription. It is likely that the hormone enhances the expression of phosphatases that act on $p70^{s6k}$, or one of its upstream regulators, to dephosphorylate these proteins. One of the questions that need to be answered is how glucocorticoids affect downstream effects of $p70^{s6k}$.

In T lymhocytes stimulation of the CD28 co-receptor causes activation of $p70^{s6k}$ and the transcription factor c-Rel (Baeuerle and Henkel 1994; Pai et al. 1994). CD28 cross-linking leads to rapid phosphorylation and degradation of IκBα, the inhibitor protein of c-Rel. Degradation of IκBα, and hence c-Rel activation, is blocked by the immunosup-

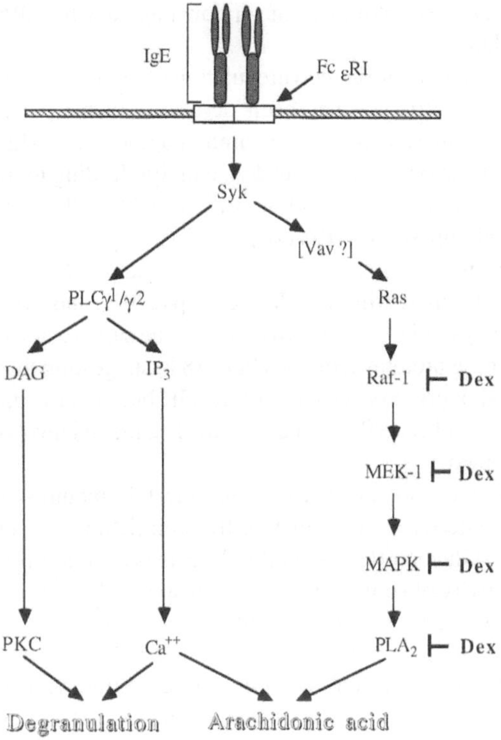

Fig. 3. Pathways for mediating degranulation of mast cells via protein kinase C (*PKC*) and Ca^{2+} (*Ca^{++}*) and release of arachidonic acids via mitogen-activated protein kinase (*MAPK*)/phospholipase A_2 (*PLA_2*). The steps shown to be inhibited by the glucocorticoid receptor have been indicated as *Dex* (the synthetic glucocorticoid dexamethasone). Activation of mitogen-activated protein kinase kinase (*MEK-1*) and Raf-1 is also inhibited by dexamethasone. *IgE*, immunoglobulin E; *DAG*, diacylglycerol; *IP_3*, inositol-(1,4,5) trisphosphate

pressant rapamycin (Lai and Tan 1994). Since $p70^{s6k}$ is rapamycin-sensitive (Chou and Blenis 1995) these results suggest that c-Rel-dependent activation of IL-2 is under the control of $p70^{s6k}$. Downregulation of $p70^{s6k}$ kinase activity by dexamethasone can therefore indirectly lead to inhibition of IL-2. This means that glucocorticoids can inhibit IL-2 production without having to act at the level of the transcription factors that control expression of this gene or downregulate the mRNA stability of this cytokine.

Another example of the downregulation of signal transduction pathways by glucocorticoids has been described in mast cells. Antigen stimulation of mast cells via the immunoglobulin E (IgE) receptor recruits cytosolic tyrosine kinases Lyn and Syk leading to the phosphorylation of several proteins and activation of PLC/PKC and the MAP kinase/phospholipase A_2 (PLA_2) pathways (Fig. 3). Glucocorticoids suppress phosphorylation of several proteins such as Raf-1, mitogen-activated protein kinase kinase 1 (MEK-1), $p42^{mapk}$, and cytosolic PLA_2 in this pathway (Rider et al. 1996) (Fig. 3). The suppression of phosphorylation of these proteins requires about 18 h, suggesting that it does not occur through a primary response of the GR but rather is dependent on prior protein synthesis (Rider et al. 1996). The factors involved have not been determined.

It is, however, not in all cases that the GR requires new protein synthesis to interfere with signaling by proinflammatory cytokines or chemokines. Glucocorticoids inhibit TNF-α production by downregulating the activity of some of the kinases responsible for the synthesis of this factor in a rapid manner (Swantek et al. 1997). The MAP kinase family members' extracellular signal-regulated kinases 1 and 2 (ERK-1 and ERK-2), p38 and Jun N-terminal kinase/stress-activated protein kinase (JNK/SAPK), and the upstream kinases MEK2, MEK3, and MEK6 are all required for TNFα gene expression. Dexamethasone inhibits lipopolysaccharide (LPS) induction of JNK/SAPK activity but not the activation of ERK-1 and -2, p38, or MEK-3, -4, and -6. The effect of dexamethasone is rapid, causing an inhibition of LPS-induced JNK/SAPK activity in 15 min (Swantek et al. 1997). This inhibition does not involve new protein synthesis and demonstrates a novel nongenomic action of the GR in antiinflammatory processes.

Rapid effects of the GR in inhibiting the Jun amino-terminal kinase pathway have been described (Caelles et al. 1997). Since the activity of c-Jun, the major component of the transcription factor AP-1, is potentiated by amino terminal phosphorylation of serine 63 and 73, inhibition of these kinases can downregulate AP-1 activity without the receptor interacting directly with this transcription factor.

Together these results show that glucocorticoids use diverse mechanisms to inhibit inflammatory processes.

4.5 Positive Regulation and Antiinflammatory Action of Glucocorticoids

Many investigators have questioned whether negative regulatory function of the GR is the only way whereby glucocorticoids exert their antiinflammatory actions. This seems not to be the case as positive action of the GR has also been implicated in several antiinflammatory processes. Below are a few examples of positive regulation of gene expression by the GR in antiinflammatory processes.

One of the typical examples is lipocortin-1, a potent inhibitor of the production of lipid inflammatory mediators and extravasation of polymorphonuclear leukocytes in inflammatory reaction (Solito et al. 1991; Flower and Rothwell 1994). However, as the contribution of lipocortin to antiinflammatory processes is controversial, I will focus on other gene products such as 5-lipoxygenase activating protein, secretory leukocyte protease inhibitor (SLPI), and the acute-phase proteins that show enhanced expression during inflammation but which are positively regulated by glucocorticoids.

4.5.1 The 5-Lipoxygenase Pathway

The 5-lipoxygenase activating protein (5-LAP or FLAP) and the enzyme 5-lipoxygenase catalyze the first two steps in the conversion of arachidonic acid to leukotrienes (McMillan and Walker 1992; Samuelsson 1993), resulting in the formation of an unstable epoxide intermediate, leukotriene A_4 (LTA_4). LTA_4, is subsequently metabolized into either LTB_4 by LTA_4 hydrolase or into LTC_4 by LTA_4 synthase. LTC_4 is converted into LTD_4 and LTE_4 by successive elimination of γ-glutamyl residue and glycine (Lewis et al. 1990; Samuelsson 1993). LTC_4, -D_4, and -E_4 are potent bronchoconstrictors that increase vascular permeability in postcapillary venules and stimulate mucus secretion. LTB_4 causes adhesion and chemotactic movement of leukocytes and stimulate aggregation, enzyme release, and generation of superoxide in neutrophils.

Leukotrienes have proinflammatory effects and have been considered as targets of the antiinflammatory action of glucocorticoids (Samuelsson 1993). However, conflicting results have been obtained on how glucocorticoids regulate their expression. It is widely accepted that in the presence of glucocorticoids, the 5-lipoxygenase pathway is repressed, but recent studies have proved otherwise. The glucocorticoid dexamethasone increases calcium ionophore-stimulated release of 5-lipoxygenase products in the human monocytic cell line, THP-1, in a dose-dependent fashion (Riddick et al. 1997). Glucocorticoids also increased immunoreactive protein and steady-state mRNA-encoding 5-lipoxygenase and its activating protein FLAP (Riddick et al. 1997). A putative consensus sequence for GR binding is present in the promoter region of the FLAP gene (Kennedy et al. 1991) and may possibly contribute to the hormone regulation of this gene.

To have a better understanding of the contribution of this pathway to inflammatory processes, mice that are genetically deficient in FLAP and 5-lipoxygenase genes have been created. FLAP-negative mice were phenotypically normal but leukotriene production was absent, confirming the significance of FLAP in the 5-lipoxygenase pathway (Griffiths et al. 1997). The defect in leukotriene production resulted in a decreased plasma protein extravasation in the peritoneal cavity in response to injection of zymosan and collagen-induced arthritis was substantially reduced in severity compared to wild-type or heterologous animals (Griffiths et al. 1997). These results together showed that FLAP plays an important role in acute and chronic inflammatory responses.

On the other hand, 5-lipoxygenase knock-out mice showed a selective opposition to inflammatory insults. They were unable to synthesize detectable levels of leukotrienes and were more resistant to lethal anaphylaxis induced by platelet-activating factor (Chen et al. 1994; Goulet et al. 1994). Reaction to ear inflammation induced by phorbol ester was normal, while inflammation induced by arachidonic acid was markedly reduced (Chen et al. 1994). These results point to a selective involvement of 5-lipoxygenase products in defined inflammatory processes. The knock-out experiments, however, do not provide a mechanistic explanation as to why glucocorticoids should enhance the activity of 5-lipoxygenase and FLAP in antiinflammatory process.

4.5.2 Secretory Leukocyte Protease Inhibitor

Glucocorticoids enhance the synthesis of SLPI, an antiprotease upregulated in human epithelial cells by inflammatory stimuli (Abbinante-Nissen et al. 1995; Stockley et al. 1986). Airway inflammatory diseases such as bronchitis, asthma, and cystic fibrosis are associated with an increase in leukocytes and their products at the sites of inflammation (Berger 1991). These leukocyte products consist of elastase, cathepsin G, and chymase which, if not regulated, could cause damage to healthy lung tissue. The regulation of the level of these proteases is brought about by the antiprotease SLPI. This provides a biological basis for the clinical observation that SLPI levels are increased in serum and respiratory lining fluid in inflammatory lung disorders (Maruyama et al. 1994; Kida et al. 1992; Fryksmark et al. 1984; Ohlsson et al. 1992). The enhanced synthesis of SLPI by glucocorticoids is unexpected and the mechanism by which this occurs is equally unclear. Glucocorticoid-induced expression, however, is sensitive to the protein synthesis inhibitor cyclohexihide indicating that it does not occur through a primary GR action (Abbinante-Nissen et al. 1995).

4.5.3 The Acute-Phase Response

Inflammation is a physiological response to a variety of stimuli such as infections and tissue injuries and it is usually accompanied by an alteration in the composition of plasma proteins in a reaction termed the acute-phase response (Kushner 1982).

The changes in protein levels that lead to the acute-phase response could either be positive or negative. Proteins whose concentrations are increased during acute inflammation are termed positive acute-phase proteins and they include haptoglobin, hemopexin, C-reactive protein, and serum amyloid A. Proteins that undergo a negative regulation are known as negative acute-phase proteins and they include albumin and transthyretin (Akira and Kishimoto 1992). Most of the proteins involved in the acute-phase response are of hepatic origin, although they can be found in other cell types such as in monocytes, macrophages, and fibroblasts. The regulated changes in protein synthesis are controlled by several cytokines, including IL-1, IL-6, and TNFα. These cytokines enhance or suppress the synthesis of the acute-phase proteins at the transcriptional level. Two of the acute-phase proteins, α1-acid glycoprotein and α2-macroglobulin, are synthesized in response to stimulation by glucocorticoids (Baumann et al. 1983; Hattori et al. 1990). In the regulation of the α1-acid glycoprotein gene, the GR has been shown to function synergistically with c/EBPβ and this is thought to occur through protein–protein interaction (Alam et al. 1993; Nishio et al. 1993). The mechanisms involved are not clear and it is equally unclear why the GR should be involved in positive regulation of the acute-phase response.

4.6 Conclusions and Perspectives

In this article, I have presented several arguments to show that although glucocorticoid-mediated downregulation of transcriptional activity of proinflammatory cytokines and chemokine genes contributes to antiinflammatory processes, many regulatory activities of the receptor are equally important.

Firstly, I have shown that in a number of cases glucocorticoid-mediated downregulation of the synthesis of proinflammatory cytokines and chemokines occurs post-transcriptionally through mRNA destabilization or through protein degradation. Unlike transcriptional repression, details of these post-transcriptional regulations are not known.

Secondly, I have discussed glucocorticoid-mediated downregulation of synthesis or action of proinflammatory cytokines at the level of cellular signaling. A number of cytokines and chemokines function in inflammation by activating a cascade of kinases and these processes can be abrogated by the GR through mechanisms that are as yet unclear.

From these considerations, it is clear that a number of questions are still open on the molecular mechanisms of antiinflammatory action of glucocorticoids. As different types of inflammatory reactions (e.g., rheumatoid arthritis, asthma, or colitis) activate different molecules, the GR must use more than one mechanism to inhibit these processes. For a broader understanding of the antiinflammatory action of glucocorticoids, these other mechanisms need to be studied to help decide where they are relevant and how they can be effectively opposed.

I have also discussed the question of whether negative or positive regulation by the GR contributes to antiinflammatory action of glucocorticoids. The current ideas are that the negative effects of the GR are important for the antiinflammatory action of glucocorticoids, while the positive effects contribute to the adverse effects that occur after prolonged treatment. For this reason various studies are in progress in several laboratories to separate these two activities of the GR. If both positive and negative regulation are involved in the antiinflammatory action by the GR, it will be very difficult for true dissociated steroids to ever be obtained. It is however possible that certain types of inflammatory disorders may need to be counteracted by the negative regulatory action of the GR or by the positive, or even a mixture of the two activities. In such cases it is useful to know which processes are affected by which actions of the GR.

Taken together, it appears that we are just at the early stages of understanding the molecular basis of the antiinflammatory action of glucocorticoids and a lot more is needed, in particular to be able to separate the antiinflammatory actions of glucocorticoids from their adverse side effects.

References

Abbinante-Nissen JM, Simpson LG, Leikauf GD (1995) Corticosteroids increase secretory leukocytes protease inhibitor transcript levels in airway epithelial cells. Am J Physiol 268:L601–L606

Akira S, Kishimoto T (1992) IL-6 and NF-IL6 in acute-phase response and viral infection. Immunol Rev 127:25–50

Alam T, An MR, Mifflin RC, Hsieh C-C, Ge X, Papaconstantinou J (1993) trans-activation of the α1-acid glycoprotein gene acute phase responsive element by multiple isoforms of C/EBP and glucocorticoid receptor. J Biol Chem 268:15681–15688

Amano Y, Lee SW, Allison AC (1993) Inhibition by glucocorticoids of the formation of interleukin-1α, interleukin-1β and interleukin 6: mediation by decreased mRNA stability. Mol Pharmacol 43:176–182

Bacus E, Andris F, Dubois PM, Urbain J, Leo O (1996) Dexamethasone inhibits the early steps of antigen receptor signaling in activated T lymphocytes. J Immunol 156:4555–4561

Baeuerle OA, Henkel T (1994) Function and activation of NFkB in the immune system. Annu Rev Immunol 12:141–179

Baumann H, Firestone GL, Burgess TL, Grosse KW, Yamamoto KR, Held WA (1983) Dexamethasone regulation of α1-acid glycoprotein and the other acute phase reactants in rat liver and hepatoma cells. J Biol Chem 253:563–570

Berger M (1991) Inflammation in the lung in cystic fibrosis. A vicious cycle that does more harm than good? Clin Rev Allergy 9:119–142

Beutler B, Krochin N, Milsark IW, Luedke C, Cerami A (1986) Control of cachectin (tumor necrosis factor) synthesis: mechanisms of endotoxin resistance. Science 232:977–980

Boumpas DT, Anastassion ED, Older SA, Tsokos GC, Nelson DL, Balow JE (1991) Dexmethasone inhibits human interleukin 2 but not interleukin 2 receptor gene expression in vitro at the level of nuclear transcription. J Clin Invest 87:1739–1747

Caelles C, Gonzáles-Sancho J, Muñoz A (1997) Nuclear hormone receptor antagonism with AP-1 by inhibition of the JNK pathway. Genes Dev 11:3351–3364

Caldenhoven E, Liden J, Wissink S, van de Solpe A, Raaijmakers J, Koenderman L, Okret S, Gustafsson J-α, van der Saag PT (1995) Negative crosstalk between Rel A and the glucocorticoid receptor: a possible mechanism for the antiinflammatory action of glucocorticoids. Mol Endocrinol 9:401–412

Cato ACB, Wade E (1996) Molecular mechanisms of anti-inflammatory action of glucocorticoids. Bioessays 18:371–378

Celada A, McKercher S, Maki RA (1993) Repression of major histocompatibility complex 1 A expression by glucocorticoids: the glucocorticoid receptor inhibits the DNA binding of the X box DNA binding protein. J Exp Med 177:691–698

Chen X-S, Sheller JR, Johnson EN, Funk CD (1994) Role of leukotrienes revealed by targeted disruption of the 5-lipoxygenase gene. Nature 372:179–182

Chou MM, Blenis J (1995) The 70-kDa S6 kinase: regulation of a kinase with multiple roles in mitogenic signaling. Curr Opin Cell Biol 7:806–814

Chuvpilo S, Schomberg C, Gerwig R, Heinfling A, Reeves R, Grummt F, Serfling E (1993) Multiple closely-linked NFAT/octamer and HMGI(Y) binding sites are part of the interleukin-4 promoter. Nucleic Acids Res 21:5694–5704

Cronstein BN, Kimmel SC, Levin RI, Martiniuk F, Weissmann G (1992) A mechanism for the antiinflammatory effects of corticosteroids: the glucocorticoid receptor regulates leukocyte adhesion to endothelial cells and expression of endothelial-leukocyte adhesion molecule I and intercellular adhesion molecule 1. Proc Natl Acad Sci USA 89:9991–9995

De Lucas LG, Johnson DR, Whitley MZ, Collins T, Pober JS (1994) cAMP and tumor necrosis factor competitively regulate transcriptional activation through the nuclear factor binding to the c-AMP-responsive element/activating transcription factor element of the endothelial leukocyte adhesion molecule-1 (E-selectin) promoter. J Biol Chem 269:19193–19196

Diamond MI, Miner JN, Yoshinaga SK, Yamamoto KR (1990) Transcription factor interactions: selectors of positive or negative regulation from a single DNA element. Science 249:1266–1272

Flower RJ, Rothwell NJ (1994) Lipocortin-1: cellular mechanisms and clinical relevance. Trends Pharmacol Sci 15:71–76

Fryksmark U, Prellner T, Tegner H, Ohlsson K (1984) Studies on the role of antileukoprotease in respiratory tract disease. Eur J Respir Dis 65:201–209

Goulet JL, Snouwaert JN, Latour AM, Coffman TM, Koller BH (1994) Altered inflammatory responses in leukotriene-deficient mice. Proc Natl Acad Sci USA 91:12852–12856

Griffiths RJ, Smith MA, Roach ML, Stock JL, Stam EJ, Milici AJ, Scampoli DN, Eskra JD, Byrum RS, Koller BH, McNeish JD (1997) Collagen-induced arthritis is reduced in 5-lipoxygenase-activating protein-deficient mice. J Exp Med 185:1123–1129

Han J, Thompson P, Beutler B (1990) Dexamethasone and pentoxifylline inhibit endotoxin-induced cachectin/tumor necrosis factor synthesis at separate points in the signaling pathway. J Exp Med 172:391–394

Hattori M, Abraham LJ, Northemann W, Fey GH (1990) Acute-phase reaction induces a specific complex between hepatic nuclear proteins and the interleukin 6 response element of the rat α2-macroglobulin gene. Proc Natl Acad Sci USA 87:2364–2368

Heck S, Bender K, Kullmann M, Göttlicher M, Herrlich P, Cato ACB (1997) IκBα-independnet downregulation of NF-κB activity by glucocorticoids. EMBO J 16:4698–4707

Hench PS, Kendall EC, Slocumb CH, Polley HF (1949) The effect of a hormone of the adrenal cortex (17-hydroxy-11-dehydrocorticosterone: compound E) and of pituitary adrenocorticotropic hormone on rheumatoid arthritis. Proc Staff Meet Mayo Clin Proc 24:181–197

Hench PS, Kendall EC, Slocumb CH, Polley HF (1950) Effects of cortisone acetate and pituitary ACTH on rheumatoid arthritis, rheumatoid fever and certain other conditions. Arch Intern Med 85:545–666

Huhtala P, Tuuttila A, Chow LT, Lohi J, Keski-Oja J, Tryggvason K (1991) Complete structure of the human gene for 92-kDa type IV collagenase. J Biol Chem 266:16485–16490

Jonat C, Rahmsdorf HJ, Park KK, Cato ACB, Gebel S, Ponta H, Herrlich P (1990) Antitumor production and antiinflammation: downregulation of AP-1 (Fos/Jun) activity by glucocorticoid hormone. Cell 62:1182–1204

Jonat C, Stein B, Ponta H, Herrlich P, Rahmsdorf HJ (1992) Positive and negative regulation of collagenase gene expression. Matrix 1:145–155

Kennedy BP, Diehl R, Boie Y, Adam M, Dixon RF (1991) Gene characterization and promoter analysis of the human 5-lipoxygenase-activating protein (FLAP). J Biol Chem 256:8511–8516

Kida K, Mizuuchi T, Takeyama K, Hiratsuka T, Jinno K, Hosoda A, Imaizumi A, Suzuki Y (1992) Serum secretory leukoprotease inhibitor levels to diagnose pneumonia in elderly. Am Rev Respir Dis 146:1426–1429

Kleinert H, Euchenhofer C, Ihrig-Biedert I, Förstermann U (1996) Glucocorticoids inhibit the induction of nitric oxide synthase II by down-regulating cytokine-induced activity of transcription factor nuclear factor-κB. Mol Pharmacol 49:15–21

König H, Ponta H, Rahmdorf HJ, Herrlich P (1992) Interference betwen pathway-specific transcription factors: glucocorticoids antagonize phorbol ester-induced AP-1 activity without altering AP-1 site occupation in vivo. EMBO J 11:2241–2246

Koretzky GA (1997) The role of Grb2-associated proteins in T-cell activation. Immunol Today 18:401–406

Kunz D, Walker G, Eberhardt W, Pfeilschifter J (1996) Molecular mechanism of dexamethasone inhibition of nitric oxide synthase expression in interleukin 1β-stimulated mesangial cells: evidence for the involvement of transcription and posttranscriptional regulation. Proc Natl Acad Sci USA 93:255–259

Kushner I (1982) The phenomenon of the acute phase response. Ann NY Acad Sci 389:39–48

Kwon OJ, Jose PJ, Robbins RA, Schall TJ, Williams PJ (1995) Glucocorticoid inhibition of RANTES expression in human lung epithelial cells. Am J Respir Cell Mol Biol 12:488–496

Lai J-H, Tan T-H (1994) CD28 signaling causes a sustained downregulation of IκBα which can be prevented by the immunosuppressant mycin. J Biol Chem 269:30077–30080

Landemarco MF, McQuillan JJ, Rosen GD, Dean DC (1992) Characterization of the promoter for vascular cell adhesion molecule-1 (VCAM-1). J Cell Biol 267:16323–16329

Lee SW, Tsou A-P, Chan H, Thomas J, Petrie K, Eugui EM, Allison AC (1988) Glucocorticoids selectively inhibit the transcription of the interleukin 1β gene and decrease the stability of interleukin 1β mRNA. Proc Natl Acad Sci USA 85:1204–1208

Lewis H, Kaszubska W, Delamarter JF, Whelan J (1994) Cooperativity between the two NF-κB complexes, mediated by high-mobility group protein I(Y) is essential for cytokine-induced expression of E-selectin promoter. Mol Cell Biol 14:5701–5709

Lewis RA, Austen KF, Soberman RJ (1990) Leukotrienes and other products of the 5-lipoxygenase pathway. N Engl J Med 323:645–655

Lukert BP, Raisz LG (1990) Glucocorticoid-induced osteoporosis: pathogenesis and management. Ann Intern Med 112:352–364

Mason HL, Myers CS, Kendall EC (1936a) The chemistry of crystalline substance isolated from the suprarenal gland. J Biol Chem 114:613–631

Mason HL, Myers CS, Kendall EC (1936b) Chemical studies of the suprarenal cortex. The identification of a substance which possesses the qualitative action of cortin. J Biol Chem 116:267–276

Maruyama M, Hay JG, Yoshimura K, Chu C-S, Crystal RG (1994) Modulation of secretory leukoprotease inhibitor gene expression in human brochial epithelial cells by phobol ester. J Clin Invest 94:368–375

McMillan RM, Walker ERH (1992) Designing therapeutically effective 5-lipoxygenase inhibitors. Trends Pharmacol Sci 13:323–330

Mitchell JA, Belvisi MG, Akarasereenont P, Robbins RA, Jung-Kwon O, Croxtall J, Barnes PJ, Vane JR (1994) Induction of cyclo-oxygenase-2 by cytokines in human epithelial cells: regulation by dexamethasone. Br J Pharmacol 113:1008–1014

Moncada S, Higgs A (1993) The l-arginine-nitric oxide pathway. N Engl J Med 329:2002–2012

Monfar M, Blenis J (1996) Inhibition of p70/p85 S6 kinase activities in T cells by dexamethasone. Mol Endocrinol 10:1107–1115

Mori A, Kaminuma O, Suko M, Inoue S, Ohmura T, Hoshino A, Asakura Y, Miyazawa K, Yokota T, Okumura Y, Okudaira H (1997) Two distinct pathways of interleukin-5 synthesis in allergen-specific human T-cell clones are suppressed by glucocorticoids. Blood 89:2891–2900

Mukaida N, Morita M, Ishikawa Y, Rice N, Okamoto S, Kasahara T, Matsushima K (1994) Novel mechanism of glucocorticoid-mediated gene repression. J Biol Chem 269:13289–13295

Nathan C (1992) Nitric oxide as a secretory product of mammalian cells. FASEB J 6:3051–3064

Nelson PJ, Ortiz BD, Pattison JM, Krensky AM (1996) Identification of a novel regulatory region critical for expression of the RANTES chemokine in activated T lymphocytes. J Immunol 157:1139–1148

Neurath MF, Pettersson S, Meyer zum Büschenfelde K-H, Stober W (1996) Local administration of antisense phosphorothioate oligonucleotides to the p65 subunit of NF-κB abrogates established experimental colitis in mice. Nat Med 2:998–1004

Nishio Y, Isshiki H, Kishimoto T, Akira S (1993) A nuclear factor for interleukin-6 expression (NF-IL6) and the glucocorticoid receptor synergistically activate transcription of the rat α1-acid glycoprotein gene via direct protein–protein interaction. Mol Cell Biol 13:1854–1862

O'Banion MK, Winn VD, Young DA (1992) cDNA cloning and functional activity of a glucocorticoid-regulated inflammatory cyclooxygenase. Proc Natl Acad Sci USA 89:4888–4892

Ohlsson K, Sveger T, Svenningson N (1992) Protease inhibitors in bronchoalveolar lavage fluid from neonates with special reference to secretory leukocyte protease inhibitor. Acta Paediatr 81:757–759

Ohtsuka T, Kubota A, Hirano T, Watanabe K, Yoshida H, Tsurutuji M, Iizuka Y, Konishi K, Tsurufuji S (1996) Glucocorticoid-mediated gene suppression of rat cytokine-induced neutrophil chemoattractant CINC/gro, a member of the interleukin-8 family, through impairment of NF-κB activation. J Biol Chem 272:1651–1659

Oikarinen A, Kylmäniemi M, Autio-Harmainen H, Autio P, Salo T (1993) Demonstration of 72-kDa and 92-kDa forms of type IV collagenase in human skin: variable expression in various blistering diseases, induction during re-epithelialization, and decrease by topical glucocorticoids. J Invest Dermatol 101:205–210

Okamoto S, Mukaida N, Yasumoto K, Rice N, Ishikawa Y, Horiguchi H, Marakami S, Matsushima K (1994) The interleukin-8 AP-1 and κB-like sites are genetic end targets of FK506-sensitive pathway accompanied by calcium mobilization. J Biol Chem 269:8582–8589

Ortiz BD, Krensky AM, Nelson PJ (1996) Kinetics of transcription factors regulating the RANTES chemokine gene reveal a developmental switch in nuclear events during T-lymphocyte maturation. Mol Cell Biol 16:202–210

Pai S-Y, Calvo V, Wood M, Bierer BE (1994) Cross-linking CD28 leads to activation of 70-kDa S6 kinase. Eur J Immunol 24:2364–2368

Paliogianni F, Ahuja SS, Balow JP, Balow JE, Boumpas DT (1993a) Novel mechanism for inhibition of human T cells by glucocorticoids. Glucocorticoids inhibit signal transduction through IL-2 receptor. J Immunol 151:4081–4089

Paliogianni F, Raptis A, Ahuja SS, Najjar SM, Boumpas DT (1993b) Negative transcriptional regulation of human interleukin 2 (IL-2) gene by glucocorticoids through interference with nuclear transcription factors AP-1 and NF-AT. J Clin Invest 91:1481–1489

Pierce JW, Schoenleber R, Jesmok G, Best J, Moore SA, Collins T, Gerritsen ME (1997) Novel inhibitors of cytokine-induced IκBα phosphorylation and endothelial cell adhesion molecule expression show anti-inflammatory effect in vivo. J Biol Chem 272:21096–21103

Piper JM, Ray WA, Daugherty MS, Griffin MR (1991) Corticosteroid use and peptic ulcer disease: role of nonsteroidal anti-inflammatory drugs. Ann Intern Med 114:735–740

Pitzalis C, Pipitone N, Bajocchi G, Hall M, Goulding N, Lee A, Kingsley G, Lanchbury J, Panayi G (1997) Corticosteroids inhibit lymphocyte binding to endothelium and intercellular adhesion. An additional mechanism for their anti-inflammatory and immunosuppressive effect. J Immunol 158:5007–5016

Radomski NW, Palmer RMJ, Mancada S (1990) Glucocorticoids inhibit the expression of an inducible, but not the constitutive, nitric oxide synthase in vascular endothelial cells. Proc Natl Acad Sci USA 87:10043–10047

Ray A, Prefontaine KE (1994) Physical association and functional antagonism between the p65 subunit of transcription factor NF-κB and the glucocorticoid receptor. Proc Natl Acad Sci USA 91:752–756

Ray A, Laforge KS, Sehgal PB (1991) Repressor to activator switch by mutations in the first Zn finger of the glucocorticoid receptor: is direct DNA binding necessary? Proc Natl Acad Sci USA 88:7086–7090

Rao A, Luo C, Hogan PG (1997) Transcription factors of the NFAT family: regulation and function. Annu Rev Immunol 15:707–747

Riddick CA, Ring WL, Baker JR, Hodulik CR, Bigby TD (1997) Dexamethasone increases expression of 5-lipoxygenase and its activating protein in human monocytes and THP-1 cells. Eur J Biochem 246:112–118

Rider LG, Hirasawa N, Santini F, Beaven MA (1996) Activation of the mitogen-activated protein kinase cascade is suppressed by low concentrations of dexamethasone in mast cells. J Immunol 157:2374–2380

Samuelsson B (1993) Leukotrienes: mediators of immediate hypersensitivity reactions and inflammation. Science 220:568–575

Scheinman RI, Gualberto A, Jewell CM, Cidlowski JA, Baldwin AS Jr (1995) Characterization of mechanisms involved in transrepression of NF-κB by activated glucocorticoid receptors. Mol Cell Biol 15:943–953

Schindler U, Baichwal VR (1994) Three NF-κB binding sites in the human E-Selectin gene required for maximal tumor necrosis factor alpha-induced expression. Mol Cell Biol 14:5820–5831

Simmons WW, Ungureanu-Longrois D, Smith GK, Smith TW, Kelley RA (1996) Glucocorticoids regulate inducible nitric oxide synthase by inhibiting tetrahydrobiopterin synthesis and l-arginine transport. J Biol Chem 271:23928–23937

Solito E, Raugei G, Meli M, Parenti L (1991) Dexamethasone induces the expression of the mRNA of lipocortin 1 and 2 and the release of lipocortin 1 and 5 in differentiated, but not undifferentiated U-937 cells. FEBS Lett 291:238–244

Stein B, Yang MX (1995) Repression of the interleukin-6 promoter by estrogen receptor is mediated by NF-κB and c/EBPβ. Mol Cell Biol 15:4971–4979

Stockley RA, Morrison HM, Kramps JA, Dijkman JH, Burnett D (1986) Elastase inhibitors of sputum sol phase: variability, relationship to neutrophil elastase inhibition and effect of corticosteroid treatment. Thorax 41:442–447

Swantek JL, Cobb MH, Geppert TD (1997) Jun N-terminal kinase/stress activated protein kinase (JNK/SAPK) is required for lipopolysaccharide stimulation of tumor necrosis factor alpha (TNF-α) translation: glucocorticoid inhibits TNF-α translation by blocking JNK/SAPK. Mol Cell Biol 17:6274–6282

Szabo SJ, Gold JS, Murphy TL, Murphy KM (1993) Identification of cis-acting regulatory elements controlling interleukin-4 gene expression in T cells: role for NF-Y and NF-AT_C. Mol Cell Biol 13:4793–4805

Thanos D, Maniatis T (1992) The high mobility group protein HMGI(Y) is required for NF-κB-dependent virus induction of the IFN-β gene. Cell 71:777–789

Tobler A, Meier R, Seitz M, Dewald B, Baggiolini M, Fey MF (1992) Glucocorticoids downregulate gene expression of GM-CSF, NAP-1/IL-8 and IL-6, but not of M-CSF in human fibroblasts. Blood 79:45–51

Vacca A, Felli MP, Farina AR, Martinotti S, Maroder M, Screpanti I, Meco D, Petrangeli E, Frati L, Gulino A (1992) Glucocorticoid receptor-mediated suppression of the interleukin 2 gene expression through impairment of the cooperativity between nuclear factor of activated T cells and AP-1 enhancer elements. J Exp Med 175:637–646

van de Solpe A, Caldenhoven E, Stade BG, Koenderman L, Raaijmakers JAM, Johnson JP, van der Saag PT (1994) 12-O-Tetradecanoylphorbol-13-acetate- and tumor necrosis factor-α-mediated induction of intercellular adhesion molecule-1 is inhibited by dexamethasone. Functional analysis of the human adhesion molecule-1 promoter. J Biol Chem 269:6185–6192

Vayssière BM, Dupont S, Choquart A, Petit F, Garcia T, Marchandeau C, Gronemeyer H, Resche-Rigon M (1997) Synthetic glucocorticoids that dissociate transactivation and AP-1 transrepression exhibit antiinflammatory activity in vivo. Mol Endocrinol 11:1245–1255

Walker G, Pfeilschifter J, Kunz D (1997) Mechanisms of suppression of inducible nitric-oxide synthase (iNOS) expression in interferon (IFN)-γ-stimulated RAW 264.7 cells by dexamethasone. Evidence for the glucocorticoid-induced degradation of iNOS protein by calpain as a key step in post-transcriptional regulation. J Biol Chem 272:16679–16687

Whitley MZ, Thanos D, Read MA, Maniatis T, Collins T (1994) A striking similarity in the organization of the E-selectin and the beta interferon gene promoters. Mol Cell Biol 14:6464–6475

Wingett D, Forcier K, Nielson CP (1996) Glucocorticoid-mediated inhibition of RANTES expression in human T lymphocytes. FEBS Lett 398:308–311

Xie Q-W, Kashiwabara Y, Nathan C (1994) Role of transcription factor NF-κB/Rel in induction of nitric oxide synthase. J Biol Chem 269:4705–4708

Yamamoto K, Arakawa T, Ueda N, Yamamoto S (1995) Transcriptional role of nuclear factor κB and nuclear factor-interleukin-6 in the tumor necrosis factor α-dependent induction of cycloxygenase-2 in MC3T3-E1 cells. J Biol Chem 270:31315–31320

Yang-Yen H-F, Chambard J-C, Sun Y-L, Smeal T, Schmidt TJ, Drouin J, Karin M (1990) Transcriptional interference between c-Jun and the glucocorticoid receptor: mutual inhibition of DNA binding due to direct protein-protein interaction. Cell 62:1205–1215

5 Activation of Progesterone and Androgen Receptors by Signal Transduction Pathways

N.L. Weigel, L.V. Nazareth, M.-C. Keightley, and Y. Zhang

5.1 Introduction

Progesterone receptors (PR) and androgen receptors (AR) are members of the steroid/thyroid hormone superfamily of ligand-activated transcription factors (Evans 1988). In common with other members of the family, they contain carboxyl terminal hormone-binding domains, amino terminal regions that are important for transcriptional activation, and DNA-binding domains that are located between the hormone-binding domain and transactivation region (Fawell et al. 1990; Tsai and O'Malley 1994). Although these receptors are activated by their cognate ligands, recent studies have shown that PR and AR also respond to signal transduction pathways by enhancing their ligand-dependent response; in some cases, they are activated in the absence of hormone. The ability to respond to alternate signaling pathways in the absence of hormone is not a uniform characteristic of steroid receptor family members. All of the estrogen receptors tested appear to be responsive under some conditions (Smith et al. 1993; Ignar-Trowbridge et al. 1992; Aronica and Katzenellenbogen 1991). However, whereas chicken (Denner et al. 1990b) and rodent (Turgeon and Waring 1994) PR are responsive, the human PR is generally not responsive (Beck et al. 1992). In the case of the androgen receptor, it appears that the human receptor will respond to specific activation pathways (Culig et al. 1994; Nazareth and Weigel 1996), but the rat receptor is unresponsive (Ikonen et al. 1994; Reinikainen et al. 1996). Finally, the glucocorticoid receptor requires a ligand for activation (Nordeen et al. 1993). The mechanism by which these receptors can be activated in the absence of ligand is a topic of active research.

5.2 Progesterone Receptors

Unlike most steroid receptors, the PR is expressed as two forms, A and B, that are both derived from the same gene either from separate mRNAs (Kastner et al. 1990) or by alternate initiation of translation at an internal start site (Conneely et al. 1989). The A form of the receptor lacks the most amino-terminal portion of the B form 128 amino acids in the case of the chicken receptor (Conneely et al. 1989) and 164 amino acids in the case of the human receptor (Kastner et al. 1990), but shares the remaining sequence. The rodent receptors are also expressed as two forms (Ilenchuk and Walters 1987), but the rabbit only expresses the B form (Loosfelt et al. 1984).

5.2.1 Chicken Progesterone Receptor

5.2.1.1 Activation Pathways

The first compelling evidence that receptors can be activated in the absence of their ligands came from the studies of Denner et al. who

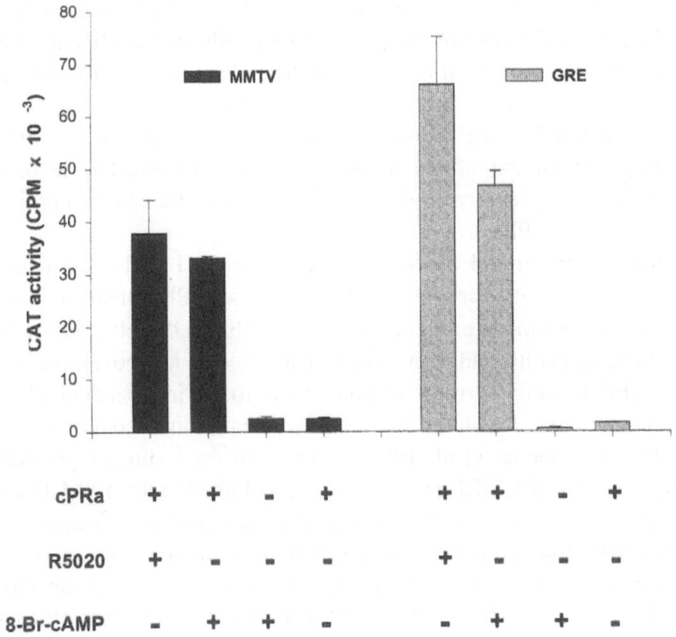

Fig. 1. Treatment of CV1 cells cotransfected with chicken progesterone receptor A (*cPR*$_A$) and mouse mammary tumor virus chloramphenicol acethyltransferase (*MMTV CAT*) with 8-Bromo cyclic adenosine monophosphate (*8-Br cAMP*) induces PR-dependent transcription. CV1 cells (2×10^5/well) were transfected with 6 ng of cPR$_A$ expression plasmid and either 0.35 μg MMTV CAT or 0.5 μg $GRE_2E1bCAT$ using the adenovirus-mediated transfection procedure (Nazareth and Weigel 1996) at 500 virus particles/cell. Cells were treated with 10 n*M* R5020 or 1 m*M* 8-Br cAMP as indicated. Cells were harvested 48 h after transfection and assayed for CAT activity

demonstrated that chicken PR_A, transiently cotransfected into CV1 cells with a reporter plasmid, containing two progesterone response elements and a portion of the thymidine kinase promoter linked to the coding region of chloramphenicol acetyl transferase (PREtkCAT), is transcriptionally activated when the cells are treated with the protein kinase A (PKA) activator, 8-Bromo cyclic adenosine monophosphate (8-Br cAMP) (Denner et al. 1990b). The activity was absolutely dependent upon receptor expression. Interestingly, treatment with the PKA inhibitors, H8, or a cell permeable fragment of PKI, reduced both ligand-independent activation as well as hormone-dependent activation indicating that this pathway also plays a role in hormone-dependent activation. Ligand-independent activation was not limited to activation by PKA activators. Treatment of the transfected cells with okadaic acid, an inhibitor of phosphatases 1 and 2 A, also activated the chicken progesterone receptor (cPR).

Although the initial studies were done in CV1 cells, subsequent studies showed that 8-Br cAMP will also activate cPR transfected in all cell lines tested to date, including HeLa cells, COS cells, SK-N-SH neuroblastoma cells, and PC3 prostate cancer cells. Moreover, the ability of cPR to activate transcription seems to be independent of the promoter context. The initial studies utilized a promoter containing two PREtkCATs (Denner et al. 1990b). When an even simpler promoter containing two PRE/GRE response elements linked to the TATA box of the E1b gene was used, cPR still activated transcription in response to 8-Br cAMP, although the extent of activation compared to hormone-dependent activation was somewhat less than in the case of the more complex promoter (Zhang et al. 1994). Figure 1 shows that cPR can induce transcription in response to 8-Br cAMP from an even more complex promoter, the mouse mammary tumor virus long terminal repeat (MMTV LTR), which contains PREs as well as half sites. This contrasts with studies of the ER which indicate that the ability to exhibit ligand-independent activation is dependent upon cell type and promoter contexts (Ince et al. 1994).

cPR_A activation is not limited to 8-Br cAMP and okadaic acid. Subsequent studies have shown that dopamine agonists of the D1 type also activate cPR (Power et al. 1991), as does calyculin A (Zhang et al. 1994), a phosphatase inhibitor. In contrast, dopamine D2 agonists and other compounds which stimulate adenylate cyclase, including isoproteronol (a β-adrenergic receptor agonist), norepinephrine, and epinephrine, do not activate cPR (Power et al. 1991) so there is some specificity to the activation. However, treatment with the phorbol ester, tetradecanoyl 12-phorbol 12-acetate (TPA), which activates PKC, also activates cPR, as shown in Fig. 2. In addition to the finding that pathways that increase Ser/Thr phosphorylation can activate cPR, other studies have demonstrated that alterations in tyrosine phosphorylation can re-

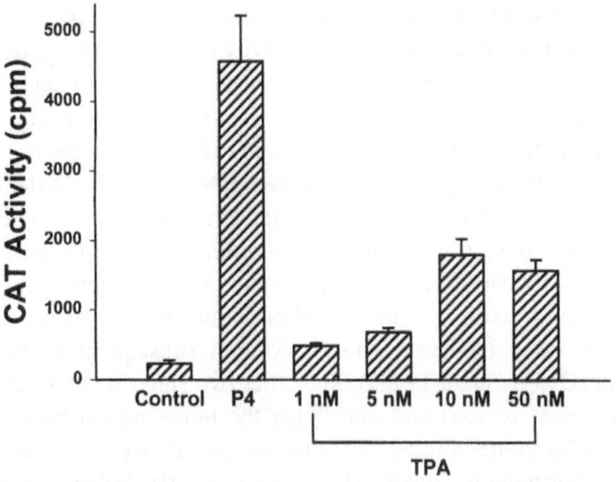

Fig. 2. Tetradecanoyl 12-phorbol 12-acetate (*TPA*) induces progesterone receptor A (PR_A)-mediated transcription. CV1 cells were cotransfected with 1 µg PR_A and 5 µg $GRE_2E1bCAT$ using polybrene as described previously (Zhang et al. 1994). Twenty-four hours after transfection, cells were treated with 10^{-7} *M* progesterone (*P4*), or with the indicated amounts of TPA. The data are presented ± standard error of mean

sult in activation of cPR. Zhang et al. have shown that treatment of transfected cells with vanadate, a phosphotyrosine phosphatase inhibitor, also causes activation of cPR (Zhang et al. 1994). Moreover, epidermal growth factor (EGF) treatment, which typically activates mitogen-activated protein (MAP) kinases, also activates cPR (Zhang et al. 1994).

The finding that so many different compounds which activate different signal transduction pathways cause activation of cPR raises the question of whether there are genuinely multiple mechanisms for activating cPR or whether these pathways converge on a single target. That there are a minimum of two pathways is indicated by the finding that replacing Ser^{628} in cPR with threonine abrogates the response to dopamine, but not to okadaic acid (Power et al. 1991).

5.2.1.2 Mechanism for Ligand-Independent Activation of cPR

The cPR is a phosphoprotein and its activity is regulated by phosphorylation (Denner et al. 1990a; Bai et al. 1994, 1997; Bai and Weigel 1996). Four phosphorylation sites have been identified in cPR using ^{32}P-labeled oviduct tissue minces as a source of radiolabeled cPR to prepare and sequence the phosphotryptic peptides (Denner et al. 1990a; Poletti and Weigel 1993). All four sites reside in the region common to PR_A and PR_B and they have been numbered according to their position in the PR_B. Two of the sites, Ser^{211} and Ser^{260} in the amino-terminal portion of the receptor, exhibit some phosphorylation in the absence of hormone, but show increased phosphorylation subsequent to hormone treatment (Denner et al. 1990a). The other two sites, Ser^{367} (also in the amino-terminal region) and Ser^{530} (in the hinge region between the DNA and hormone-binding domain) are phosphorylated only in response to hormone treatment (Denner et al. 1990a; Poletti and Weigel 1993). The sites and their hormone dependence of phosphorylation are remarkably well conserved when cPR is expressed as a heterologous protein. cPR expressed in yeast (*Saccharomyces cerevisiae*) (Poletti et al. 1993) and mammalian cells (COS and CV1) (Bai et al. 1997; Allgood et al. 1997) exhibit indistinguishable phosphorylation patterns. Mutation of either Ser^{211} or Ser^{260} to alanine substantially reduces the hormone-dependent activation of cPR (Bai and Weigel 1996; Bai et al. 1997), whereas phosphorylation of Ser^{530} allows the receptor to respond to lower levels of hormone (Bai et al. 1994).

The role of these sites in the ligand-independent activation of cPR by 8-Br cAMP was tested by comparing hormone-dependent activation with ligand-independent activation using receptors with alanine substitutions at each of the four phosphorylation sites, as well as at all four sites (Bai et al. 1997). Ligand-independent activation was substantially reduced when either Ser^{211} or Ser^{260} was mutated. However, mutation of all four phosphorylation sites did not completely eliminate the ligand-independent activation of cPR, demonstrating that phosphorylation of one or more of these sites is not the initiator of ligand-independent activation. Similar results were obtained when a dopamine agonist was used to activate the receptor. To determine whether the 8-Br cAMP treatment caused phosphorylation of a novel site, COS cells were transfected with a cPR_A expression plasmid using lysine-coupled inactivated adenovirus as a carrier to achieve high efficiency transfection, labeled with ^{32}P, receptor-isolated, and phosphopeptides were prepared and separated (Bai et al. 1997). These studies showed that there were no changes in net phosphorylation, nor were novel sites phosphorylated. Interestingly, the receptor activated by 8-Br cAMP displayed a phosphorylation pattern indistinguishable from the control receptor. Consistent with this finding were the functional studies showing that mutation of $Ser^{211}$and Ser^{260}, but not Ser^{367} or Ser^{530}, reduced the ligand-independent response. If activators of ligand-independent pathways do not alter phosphorylation of cPR, what, then, is the target of this pathway? The most likely candidates are proteins in the heat shock complex which normally maintain the receptor in an inactive state or a protein that interacts with a receptor to produce a transcriptionally active complex. One appealing target would be a coactivator that interacts with the receptor as well as with the general transcription apparatus. A preliminary report from Rowan et al. (1997) describes increased phosphorylation of the steroid receptor coactivator (SRC-1) in response to 8-Br cAMP treatment, suggesting that this coactivator may be the target of the PKA pathway.

5.2.2 Rat Progesterone Receptor

Evidence that the rat PR can be activated in the absence of hormone comes from both in vitro and in vivo studies. Using a transfected reporter plasmid to detect PR activation, Turgeon and Waring first found that rat PR in primary rat pituitary cells can be activated by 8-Br cAMP or by gonadotropin-releasing hormone (Turgeon and Waring 1994). That the response was PR-dependent was shown by the ability of the progesterone antagonist, mifepristone (RU486), to block the activation.

Using lordosis, a steroid-dependent behavioral response, as a marker of progesterone action, several studies have now shown that PR expressed in the brain can be activated by signal transduction pathways in the absence of ligand. Mani et al. (1994a) reported that ovariectomized estrogen-primed animals exhibited lordosis in response to direct administration of D_1 dopamine agonists into the third ventricle. The ability of two progesterone antagonists, RU486 and onapristone (ZK98299), to block this response indicates that the dopamine agonist action is mediated by PR. Subsequent studies in which antisense or sense oligonucleotides coding for PR were administered prior to the dopamine showed that only the antisense oligonucleotide reduced the lordosis response (Mani et al. 1994b). Subsequent studies by Apostolakis et al. (1996a) showed that the D1-like agonists, SKF77434, SKF75640, and SKF85174 were acting through a D5 dopamine receptor. These conclusions were based on the finding that antisense oligonucleotides to D5 receptors, but not to D1 receptors, blocked agonist-induced lordosis.

Additional studies showed that cocaine can induce lordosis in estrogen primed rats (Apostolakis et al. 1996b). That the cocaine acted through PR was demonstrated by inhibition of the response by RU486 or by PR antisense oligonucleotides. Similar to dopamine, cocaine acts through the D5 dopamine receptor as shown by the inhibition of the response to cocaine by antisense oligonucleotides to the D5 dopamine receptor, as well as by D1-like antagonists. Cocaine is known to inhibit a membrane-associated dopamine transporter (DAT). Inhibition of DAT using antisense oligonucleotides also induced lordosis which was blocked by pretreatment with the D1-like/D2 dopamine receptor antagonist, *N*-ethoxycarbonyl-2 ethoxy-1,2-dihydroquinoline.

5.2.3 Mouse Progesterone Receptor

The mouse also exhibits lordosis in response to dopamine treatment of estrogen-primed mice (Mani et al. 1996). To directly determine whether dopamine-induced lordosis requires PR, the recently developed PR knockout (PRKO) mice were studied. Estrogen-primed ovariectomized mice failed to exhibit lordosis in response to progesterone treatment and also did not respond to treatment with the dopamine agonist SKF 38393. The PRKO mice, however, have not completely lost their ability to respond because activation of an independent pathway, by serotonin administration, induced equivalent lordosis quotients in the wild-type and PRKO mice. These studies confirm that the dopamine response is mediated by the PR.

5.2.4 Human Progesterone Receptor

In contrast to the studies of cPR and rodent PRs, there have been very few reports of ligand-independent activation of the human PR. Kazmi et al. (1993) reported ligand-independent activation in response to 8-Br cAMP of hPR-B and hPR-A cotransfected into COS cells with a PREtkCAT reporter, but not with MMTVCAT. Activity was compared to the activity induced by a suboptimal amount of the progestin, R5020 (10^{-10} *M*), so comparisons to optimal hormone-dependent activation are not possible. Subsequently, Philpott and Shahid (1996) have demonstrated that dopamine treatment of hPR-B transfected into COS cells induces activation. The reasons for the differences between the results of these investigators and the studies described below are not clear.

Several investigators have reported that endogenous hPR in T47D cells fails to exhibit ligand-independent activation in response to 8-Br cAMP (Beck et al. 1992, 1993; Satorius et al. 1993). Nonetheless, 8-Br cAMP significantly enhances the response to R5020 (Beck et al. 1992). Weigel and Zhang demonstrated that hPR was not responsive to 8-Br cAMP under identical conditions to those that activate cPR (Weigel and Zhang 1997).

However, similar to the results with cPR, treatment with the PKA inhibitor, H8, blocked both the 8-Br cAMP effect and the hormone-induced transcription (Beck et al. 1992). Western blot analysis showed that the treatment had not reduced the expression of PR. Moreover, the activity of a constitutively active reporter, pSV_2CAT, was unaffected by H8. This, again, suggests that the PKA pathway is directly or indirectly involved in transcriptional activation of PRs.

Other modulators of kinase and phosphatase activity also stimulate the ligand-dependent activity of PR. The phosphatase inhibitor, okadaic acid, also stimulates ligand-dependent response. Western blotting showed that the increased activity in response to 8-Br cAMP or to okadaic acid was not due to increased expression of PR (Beck et al. 1992). Moreover, treatment with 8-Br cAMP, okadaic acid, or H8 did not change net phosphorylation of PR (Beck et al. 1992), suggesting that these pathways are modifying proteins that functionally interact with PR.

Additional studies have shown that the agonist activity of R5020 can also be enhanced by treatment with TPA, an activator of PKC (Edwards et al. 1993). The stimulation by TPA of hormone-dependent activity was greater than that observed using 8-Br cAMP.

Although the basis for the differences between the response of cPR and hPR has not yet been determined, in vitro studies of the DNA binding of the two PRs suggest that hPR requires ligand for DNA binding as assessed by electrophoretic mobility shift assays (EMSA) (Beck et al. 1992), whereas cPR does not (Rodriguez et al. 1989; Tsai et al. 1988). Antagonists that induce DNA binding may exhibit very weak agonist activity (Meyer et al. 1990). Beck et al. have found that co-treatment with the antagonist RU486 and 8-Br cAMP causes activation of hPR as measured by a stably integrated MMTV reporter in T47D cells, turning the activity of RU486 from that of an antagonist into that of an agonist (Beck et al. 1993). This activity was specifically induced by 8-Br cAMP as neither okadaic acid nor TPA induced a comparable response. Subsequent studies showed that 8-Br cAMP with RU486 also induced the endogenous human metallothionein-IIA gene (Edwards et al. 1993). The response to 8-Br cAMP was strictly dependent upon antagonists that induced DNA binding as judged by EMSA since ZK98299 failed to stimulate PR activity in combination with 8-Br cAMP (Beck et al. 1993).

Sartorius et al. (1993) also found that the combination of 8-Br cAMP and RU486 induced the activity of PR and subsequently showed, using T47D lines that express only PR-A or PR-B, that only the PR-B form of the receptor is activated under these conditions (Sartorius et al. 1994). More recently, Jackson et al. (1997) have isolated a protein, L7/SPA, that interacts with the hinge region of hPR and enhances the partial agonist activity of RU486. Whether this protein plays a role in the antagonist/agonist switch has not yet been reported.

Taken together, these studies show that signaling pathways modulate the activity of hPR, but that under most conditions identified to activate other PRs in the absence of ligand hPR remains inactive.

5.3 Androgen Receptors

ARs are also members of the steroid/thyroid hormone receptor family of ligand-activated transcription factors. Until recently, only one form of the AR had been described. However, there is now evidence that a smaller A-like form corresponding to the A form of the PR is also expressed (Wilson and McPhaul 1994). This form typically represents <15% of the AR and its role in AR function is not yet well characterized. In contrast to the studies of chicken and rodent PRs, reports of ligand-independent activation of ARs have been mixed.

5.3.1 Human Androgen Receptor

Culig et al. (1994) found that human androgen receptor (hAR) transiently expressed in DU145 prostate cancer cells could activate a reporter consisting of two AREs and a TATA box linked to the coding region of chloramphenicol acetyl transferase when the cells were treated with insulin-like growth factor (IGF)-I, to a lesser extent by keratinocyte growth factor (KGF), and even more weakly by EGF. In contrast, when the more complex prostate specific antigen promoter which is also androgen responsive was used, only IGF-I was capable of inducing AR-mediated ligand-independent transactivation (Culig et al. 1994). When the simplest promoter consisting of a single ARE and a TATA box was tested, again only IGF-I induced AR-dependent activation. Growth factor-mediated activation was blocked in all cases by co-administration of the antiandrogen, casodex.

To measure induction of an endogenous gene, the effects of androgens and growth factors on the induction of prostate-specific antigen (PSA), a natural target gene in LNCaP cells which contain endogenous AR, were measured (Culig et al. 1994). Cells were treated with hormone or growth factors in serum-free conditions. Under these conditions IGF-I induced PSA production; this induction was inhibited by treatment with casodex. Neither IGF-II, basic fibroblast growth factor (bFGF), KGF, nor EGF induced PSA expression (Culig et al. 1994).

Nazareth and Weigel subsequently showed that human AR, transiently transfected into CV1 cells, can be activated by treatment with forskolin, an activator of adenylate cyclase (Nazareth and Weigel 1996). This activity was blocked by the antiandrogens, casodex and flutamide. An intact DNA binding domain was required for this activation, suggesting that the activation requires DNA binding. AR, transiently expressed in PC3 prostate cancer cells that lack endogenous receptor, also induced expression of a target gene containing two AREs and a TATA

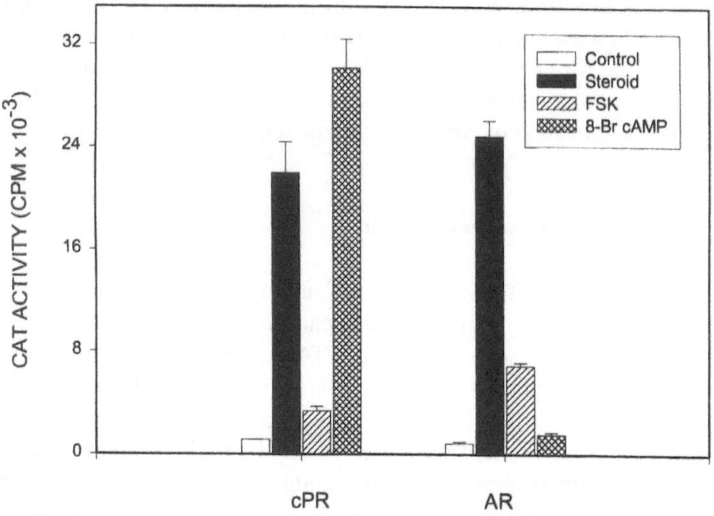

Fig. 3. Response of chicken progesterone receptor (*cPR*) and androgen receptor (*AR*) to protein kinase A modulators. CV1 cells were cotransfected with 6.25 ng cPR expression vector or 62.5 ng AR expression vector, and 0.5 μg of $GRE_2E1bCAT$ using the adenovirus-mediated transfection procedure as described previously (Nazareth and Weigel 1996). Cells were treated with ethanol, steroid [10^{-8} *M* progesterone for PR and 10^{-10} *M* R1881 (an androgen analogue) for AR], 40 μ*M* forskolin (*FSK*), or 2 m*M* 8-Bromo cyclic adenosine monophosphate (*8-Br cAMP*) as indicated. Cells were harvested after 48 h and assayed for CAT activity. The *error bars* indicate the standard error of the mean

box or a more complex promoter containing a portion of the 5' flanking region of the probasin gene when treated with forskolin. Similar to the results with PR, PKI, a specific PKA inhibitor reduced both ligand-independent and ligand-dependent activity of AR under conditions which did not alter induction of pSV_2CAT (Nazareth and Weigel 1996). This inhibition was not due to reduced expression of AR.

Finally, there is evidence that the AR in human prostate explants can be activated in the absence of androgen by PKA signaling pathways. Estradiol, bound to sex hormone-binding globulin (SHBG) induces formation of cAMP resulting in induction of PSA, an AR-responsive gene (Nakhla et al. 1997). This response is blocked by the anti-androgens, hydroxyflutamide and cyproterone acetate, intimating that the response is AR-mediated. Neither estradiol, nor SHBG alone induces PSA secretion and the response can be mimicked by administration of forskolin or cAMP. SHBG and estradiol induction can be inhibited by PKI; however, induction of the gene by dihydrotestosterone (DHT), the natural AR ligand, cannot be blocked by PKI.

Despite the apparent similarities to cPR activation, the pathways to cPR and AR activation do not appear to be identical, as shown in Fig. 3. In this experiment, CV1 cells were cotransfected with either AR or cPR_A expression plasmids and a reporter responsive to both receptors ($GRE_2E1bCAT$) using lysine-coupled adenovirus as a carrier for the plasmids and treated with hormone, 8-Br cAMP (which directly activates PKA), or forskolin (which increases cAMP levels due to increasing the activity of adenylyl cyclase). In the case of cPR, 8-Br cAMP is more effective as an activator than is forskolin, whereas the reverse is true for AR. This has been a consistent observation and suggests that the pathways leading to the activation of the two receptors must differ in some as yet unidentified way.

Analysis of the effects of forskolin treatment on AR showed that forskolin treatment caused more of the receptor to be found in the salt-extractable (nuclear) fraction when cells were fractionated (Nazareth and Weigel 1996). Moreover, the receptor in whole-cell extracts bound somewhat better to DNA in EMSA assays than did the receptor from untreated cells. Casodex treatment blocked both of these responses to forskolin.

The effect of forskolin on AR phosphorylation under conditions that elicit ligand-independent activation have not been studied in detail. However, it has been demonstrated previously that AR migrates as a doublet on sodium dodecyl sulfate (SDS) gels due to differential phosphorylation (Jenster et al. 1994). Treatment with forskolin results in more of the faster migrating form (Nazareth and Weigel 1996) characteristic of the less phosphorylated receptor than does no treatment or hormone treatment. This would suggest that AR may be dephosphorylated in response to forskolin, but the functional consequence of the change remains to be determined.

Despite these reports of ligand-independent activation, human AR stably expressed in CHO cells is not activated by 8-Br cAMP or by phorbol myristate acetate (PMA) in the absence of androgen. PMA caused a small increase in the hormone-dependent response when an MMTV reporter was used, but not when GREtkLUC or PSA-LUC reporters were used (de Ruiter et al. 1995).

Collectively, the studies of the human AR suggest that its ability to respond to signal transduction pathways is highly dependent upon the activator, promoter, and cell context.

5.3.2 Rodent Androgen Receptors

Analyses of the response of the rat AR to activators of signal transduction pathways show that ligand-dependent activation can be enhanced by 8-Br cAMP, PMA, forskolin, or okadaic acid, but that in all cases the receptor was inactive in the absence of hormone (Ikonen et al. 1994). A subsequent study showed that growth factors such as EGF and IGF-I stimulated hormone-dependent activity, but failed to induce ligand-independent activation (Reinikainen et al. 1996). The basis for the differences in response between the human and rat AR has not been determined. Although the difference in response may be strictly a species difference, other possibilities include differences in reporters or methods of transfection [relatively inefficient $CaPO_4$ transfection for rat AR hormone (Ikonen et al. 1994) versus the higher efficiency lipofectamine (Culig et al. 1994) and adenovirus-mediated procedures (Nazareth and Weigel 1996) for human androgen receptors]. Also of interest is the report of Gupta et al. on effects of androgens and epidermal growth factor on fetal mouse sexual differentiation (Gupta et al. 1996). In this system, in which 13-day mouse female reproductive tracts are cultured in vitro, cyproterone acetate blocked EGF-induced Wolffian duct stabilization. Moreover, EGF-induced proliferation of isolated reproductive tract cells was blocked by cyproterone acetate. These studies suggest that AR is activated by EGF; however, EGF also increased the expression of AR three fold, so the growth may have been due to an increased ability to respond to any trace of androgen remaining in the dialyzed serum.

5.4 Conclusion

The studies described herein demonstrate that there is substantial cross-stalk between steroid receptors and signal transduction pathways. In some cases, activation of signal transduction pathways promotes activation of steroid receptors in the absence of hormone, and there is abundant evidence that these pathways can enhance the hormone-dependent responses of receptors. That the signaling pathways and protein phosphorylation play important roles in hormone-dependent activation is also evident. Mutation of the phosphorylation sites in cPR substantially alters the activity of the receptor (Bai et al. 1994, 1997; Bai and Weigel 1996). In each case examined, inhibition of PKA inhibits the hormone-dependent activation of steroid receptors (Beck et al. 1992; Nazareth and Weigel 1996; Denner et al. 1990a). This is consistent with the reports from the Rosner laboratory showing that steroids [estradiol or androgens bound to SHBG (Nakhla and Rosner 1996), or progesterone bound to CBG] increase intracellular levels of cAMP (Nakhla et al. 1988). Consequently, in many cases, the steroid hormones serve the dual purpose of raising cAMP levels as well as directly binding to and activating their cognate steroid receptors. These studies suggest that, in vivo, both signal transduction pathways leading to altered phosphorylation of receptors and associated proteins, as well as the steroids, play important roles in the regulation of receptor activation.

References

Allgood VE, Zhang Y, O'Malley BW, Weigel NL (1997) Analysis of chicken progesterone receptor function and phosphorylation using an adenovirus mediated procedure for high efficiency DNA transfer. Biochem 36(1):224–232

Apostolakis EM, Garai J, Clark JH, O'Malley BW (1996a) In vivo regulation of central nervous system progesterone receptors: cocaine induces steroid dependent behavior through dopamine transporter modulation of D5 receptors in rats. Mol Endocrinol 10:1595–1604

Apostolakis EM, Garai J, Fox C, Smith CL, Watson SJ, Clark JH, O'Malley BW (1996b) Dopaminergic regulation of progesterone receptors: Brain D5 dopamine receptors mediate induction of lordosis by D1-like agonists in rats. J Neurosci 16(6):4823–4834

Aronica SM, Katzenellenbogen BS (1991) Progesterone receptor regulation in uterine cells: stimulation by estrogen, cyclic adenosine 3',5'-monophosphate, and insulin-like growth factor I and suppression by antiestrogens and protein kinase inhibitors. Endocrinology 128:2045–2052

Bai W, Weigel NL (1996) Phosphorylation of Ser^{211} in the chicken progesterone receptor modulates its transcriptional activity. J Biol Chem 271(22):12801–12806

Bai W, Tullos S, Weigel NL (1994) Phosphorylation of Ser^{530} facilitates hormone-dependent transcriptional activation of the chicken progesterone receptor. Mol Endocrinol 8:1465–1473

Bai W, Rowan BG, Allgood VE, O'Malley BW, Weigel NL (1997) Differential phosphorylation of chicken progesterone receptor in hormone-dependent and ligand-independent activation. J Biol Chem 272:10457–10463

Beck CA, Weigel NL, Edwards DP (1992) Effects of hormone and cellular modulators of protein phosphorylation on transcriptional activity, DNA binding, and phosphorylation of human progesterone receptors. Mol Endocrinol 6:607–620

Beck CA, Weigel NL, Moyer ML, Nordeen SK, Edwards DP (1993) The progesterone antagonist RU486 acquires agonist activity upon stimulation of cAMP signaling pathways. Proc Natl Acad Sci USA 90:4441–4445

Conneely OM, Kettelberger DM, Tsai M-J, Schrader WT, O'Malley BW (1989) The chicken progesterone receptor A and B isoforms are products of an alternate translation initiation event. J Biol Chem 264:14062–14064

Culig Z, Hobisch A, Cronauer MV, Radmayr C, Trapman J, Hittmair A, Bartsch G, Klocker H (1994) Androgen receptor activation in prostatic tumor cell lines by insulin-like growth factor-I, keratinocyte growth factor, and epidermal growth factor. Cancer Res 54:5474–5478

Denner LA, Schrader WT, O'Malley BW, Weigel NL (1990a) Hormonal regulation and identification of chicken progesterone receptor phosphorylation sites. J Biol Chem 265:16548–16555

Denner LA, Weigel NL, Maxwell BL, Schrader WT, O'Malley BW (1990b) Regulation of progesterone receptor-mediated transcription by phosphorylation. Science 250:1740–1743

de Ruiter PE, Teuwen R, Trapman J, Dijkema R, Brinkmann AO (1995) Synergism between androgens and protein kinase C on androgen-regulated gene expression. Mol Cell Endocrinol 110:R1–R6

Edwards DP, Weigel NL, Nordeen SK, Beck CA (1993) Modulators of cellular protein phosphorylation alter the trans-activation function of human progesterone receptor and the biological activity of progesterone antagonists. Breast Cancer Res Treatment 27:41–56

Evans RM (1988) The steroid and thyroid hormone receptor superfamily. Science 240:889–895

Fawell SE, Lees JA, White R, Parker MG (1990) Characterization and colocalization of steroid binding and dimerization activities in the mouse estrogen receptor. Cell 60:953–962

Gupta C, Chandorkar A, Nguyen AP (1996) Activation of androgen receptor in epidermal growth factor modulation of fetal mouse sexual differentiation. Mol Cell Endocrinol 123:89–95

Ignar-Trowbridge DM, Nelson KG, Bidwell MC, Curtis SW, Washburn TF, Machlachlan JA, Korach KS (1992) Coupling of dual signaling pathways: epidermal growth factor action involves the estrogen receptor. Proc Natl Acad Sci USA 89:4658–4662

Ikonen T, Palvimo JJ, Kallio PJ, Reinikainen P, Janne OA (1994) Stimulation of androgen-regulated transactivation by modulators of protein phosphorylation. Endocrinology 4:1359–1366

Ilenchuk TT, Walters MR (1987) Rat uterine progesterone receptor analyzed by [3H]R5020 photoaffinity labeling: evidence that the A and B subunits are not equimolar. Endocrinology 120(4):1449–1456

Ince BA, Montano MM, Katzenellenbogen BS (1994) Activation of transcriptionally inactive human estrogen receptors by cyclic adenosine 3', 5'-monophosphate and ligands including antiestrogens. Mol Endocrinol 8:1397–1406

Jackson TA, Richer JK, Bain DL, Takimoto GS, Tung L, Horwitz KB (1997) The partial agonist activity of antagonist-occupied steroid receptors is controlled by a novel hinge domain-binding coactivator L7/SPA and the corepressors N-CoR or SMRT. Mol Endocrinol 11:693–705

Jenster G, deRuiter PE, van der Korput AGM, Kuiper GGJM, Trapman J, Brinkmann AO (1994) Changes in the abundance of androgen receptor isotypes: effects of ligand treatment, glutamine-stretch variation, and mutation of putative phosphorylation sites. Biochemistry 33:14064–14072

Kastner P, Krust A, Turcotte B, Strupp U, Tora L, Gronemeyer H, Chambon P (1990) Two distinct estrogen-regulated promoters generate transcripts encoding the two functionally different human progesterone receptor forms A and B. EMBO J 9:1603–1614

Kazmi SMI, Visconti V, Plante RK, Ishaque A, Lau C (1993) Differential regulation of human progesterone receptor A and B form-mediated trans-activation by phosphorylation. Endocrinology 144:1230–1238

Loosfelt H, Logeat F, Hai MTV, Milgrom E (1984) The rabbit progesterone receptor: evidence for a single steroid-binding subunit and characterization of receptor mRNA. J Biol Chem 259:14196–14202

Mani SK, Allen JM, Clark JH, Blaustein JD, O'Malley BW (1994a) Convergent pathways for steroid hormone- and neurotransmitter-induced rat sexual behavior. Science 265:1246–1249

Mani SK, Blaustein JD, Allen JM, Law SW, O'Malley BW, Clark JH (1994b) Inhibition of rat sexual behavior by antisense oligonucleotides to the progesterone receptor. Endocrinology 135:1409–1414

Mani S, Allen JMC, Lydon JP, Mulac-Jericevic B, Blaustein JD, DeMayo FJ, Conneely OM, O'Malley BW (1996) Dopamine requires the unoccupied progesterone receptor in induce sexual behavior in mice. Mol Endocrinol 10:1728–1737

Meyer ME, Pornon A, Ji JW, Bocquel MT, Chambon P, Gronemeyer H (1990) Agonistic and antagonistic activities of RU486 on the functions of the human progesterone receptor. EMBO J 9:3923–3932

Nakhla AM, Rosner W (1996) Stimulation of prostate cancer growth by androgens and estrogens through the intermediacy of sex hormone-binding globulin. Endocrinology 137:4126–4129

Nakhla AM, Khan MS, Rosner W (1988) Induction of adenylate cyclase in a mammary carcinoma cell line by human corticosteroid-binding globulin. Biochem Biophys Res Commun 153(3):1012–1018

Nakhla AM, Romas NA, Rosner W (1997) Estradiol activates the prostate androgen receptor and prostate-specific antigen secretion through the intermediacy of sex hormone-binding globulin. J Biol Chem 272(11):6838–6841

Nazareth LV, Weigel NL (1996) Activation of the human androgen receptor through a protein kinase A signalling pathway. J Biol Chem 271(33):19900–19907

Nordeen SK, Bona BJ, Moyer ML (1993) Latent agonist activity of the steroid antagonist, RU486 is unmasked in cells treated with activators of protein Kinase A. Mol Endocrinol 7:731–742

Philpott AJ, Shahid M (1996) Dopamine-mediated activation of the human progesterone receptor. Cell Mol Neurobiol 16(3):417–420

Poletti A, Weigel NL (1993) Identification of a hormone-dependent phosphorylation site adjacent to the DNA-binding domain of the chicken progesterone receptor. Mol Endocrinol 7:241–246

Poletti A, Conneely OM, McDonnell DP, Schrader WT, O'Malley BW, Weigel NL (1993) Chicken progesterone receptor expressed in Saccharomyces cervisiae is correctly phosphorylated at all four ser-pro phosphorylation sites. Biochemistra 32:9563–9569

Power RF, Mani SK, Codina J, Conneely OM, O'Malley BW (1991) Dopaminergic and ligand-independent activation of steroid hormone receptors. Science 254:1636–1639

Reinikainen P, Palvimo JJ, Janne OA (1996) Effects of mitogens on androgen receptor-mediated transactivation. Endocrinology 137(10):4351–4357

Rodriguez R, Carson-Jurica MA, Weigel NL, O'Malley BW, Schrader WT (1989) Hormone induced changes in the in vitro DNA-binding activity of the chicken progesterone receptor. Mol Endocrinol 3:356–362

Rowan B, Weigel NL, O'Malley BW (1997) Phosphorylation of chicken progesterone receptor and steroid receptor coactivator-1 during ligand-independent activation by 8-Bromo-cAMP (abstract). The Endocrine Society: program and abstracts

Sartorius CA, Tung L, Takimoto GS, Horwitz KB (1993) Antagonist-occupied human progesterone receptors bound to DNA are functionally switched to transcriptional agonists by cAMP. J Biol Chem 268:9262–9266

Sartorius CA, Groshong SD, Miller LA, Powell RL, Tung L, Takimoto GS, Horwitz KB (1994) New T47D breast cancer cell lines for the independent study of progesterone B- and A-receptors: only antiprogestin-occupied B-receptors are switched to transcriptional agonists by cAMP. Cancer Res 54:3868–3877

Smith CL, Conneely OM, O'Malley BW (1993) Modulation of the ligand-independent activation of the human estrogen receptor by hormone and antihormone. Proc Natl Acad Sci USA 90:6120–6124

Tsai M-J, O'Malley BW; Tsai M-J, O'Malley BW (eds) (1994) MBIU: mechanism of steroid hormone regulation of gene transcription. Landes, Austin

Tsai SY, Carlstedt-Duke J, Weigel NL, Dahlman K, Gustafsson J-A, Tsai M-J, O'Malley BW (1988) Molecular interactions of steroid hormone receptor with its enhancer element: evidence for receptor dimer formation. Cell 55:361–369

Turgeon JL, Waring DW (1994) Activation of the progesterone receptor by the gonadotropin-releasing hormone self-priming signaling pathway. Mol Endocrinol 8:860–869

Weigel NL, Zhang Y (1997) Ligand-independent activation of steroid hormone receptors. J Mol Med (in press)

Wilson CM, McPhaul MJ (1994) A and B forms of the androgen receptor are present in human genital skin fibroblasts. Proc Natl Acad Sci USA 91:1234–1238

Zhang Y, Bai W, Allgood VE, Weigel NL (1994) Multiple signaling pathways activate the chicken progesterone receptor. Mol Endocrinol 8:577–584

6 Ligand-Independent Activation of the Estrogen Receptor: Activation by Epidermal Growth Factor and Dopamine

O. Donzé and D. Picard

6.1 Introduction

The estrogen receptor (ER) is a member of a large family of transcription factors including steroid/thyroid hormone receptors (Evans 1988). These receptors share a conserved structural and functional organization, which includes separable domains for hormone binding (HBD), DNA binding, and transcriptional activation (Tsai and O'Malley 1994). Like other members of the nuclear hormone receptor superfamily, ER has at least two transcriptional activation functions (AFs), one in its amino-terminal region (AF-1) and the second in its carboxy-terminal, ligand-binding region (AF-2) (Kumar et al. 1987; Webster et al. 1988). According to the classical model of steroid hormone action, hormone enters cells by passive diffusion where it binds to and induces a conformational change in its cognate receptor protein. This leads to the release of the inhibitory heat shock protein 90 complex, nuclear translocation, interaction with chromatin, and modulation of gene expression. Steroid receptors can activate transcription of genes containing in their 5' flanking region a specific DNA sequence called the hormone response element (HRE; ERE for ER). Before and upon ligand binding, ER is phosphorylated at several sites (Ali et al. 1993; Le Goff et al. 1994). The importance of phosphorylation in modulating transcriptional functions of ER, as well as of other steroid receptors, has been highlighted by correlative studies suggesting that phosphorylation may be involved in modulating either steroid binding, DNA binding or transactivation (Bagchi et al. 1992; Takimoto et al. 1992). Indeed, various agents that stimulate protein phosphorylation can induce transactivation of different steroid receptors in the absence of their cognate ligands (Denner et al. 1990; Power et al. 1991; Aronica and Katzenellenbogen 1993).

The discovery of ligand-independent activation was a big surprise since it had been an established dogma that ligand binding is an absolute requirement for activation of steroid receptors. Nevertheless, it was already well accepted that nuclear steroid receptors and second messenger signal transduction pathways affecting other transcription factors could synergize at the level of a target promoter (Miner and Yamamoto 1991; Saatcioglu et al. 1994). Not all members of the steroid receptor family, however, exhibit ligand-independent activation. For example, it has not been possible to activate the mineralocorticoid receptor, the human PR (hPR) or the glucocorticoid receptor (GR) to a significant extent in the absence of ligand. Ligand-independent activation has been well established for the chicken progesterone receptor (cPR), the estrogen receptor (ER), and the androgen receptor (AR) (Weigel and Zhang 1997). Several reagents have been found to induce ER in the absence of ligand: cyclic adenosine monophosphate (cAMP) (Aronica and Katzenellenbogen 1993; El-Tanani and Green 1997); okadaic acid (inhibitor of protein phosphatases 1 and 2 A) (Power et al. 1991); insulin (Ma et al. 1994; Newton et al. 1994; Patrone et al. 1996); dopamine (Power et al. 1991; Smith et al. 1993); epidermal growth factor (EGF) (Ignar-Trowbridge et al. 1993, 1995; Bunone et al. 1996; El-Tanani and Green 1997); insulin-like growth factor I (IGF-I) (Aronica and Katzenellenbogen 1993; Ma et al. 1994; Newton et al. 1994); transforming growth factor α (TGFα) (Ignar-Trowbridge et al. 1993, 1995);

erbB2 (Pietras et al. 1995), and phorbol ester (Ignar-Trowbridge et al. 1995, 1996; Bunone et al. 1996; Patrone et al. 1996).

6.2 Activation of the Estrogen Receptor by Epidermal Growth Factor

We have focused mainly on the ligand-independent activation of the human ER (hER) using a variety of assays in different transfected cell lines (SK-Br-3, COS-1, Hela, CV1) (Bunone et al. 1996; Picard et al. 1997). With a transcriptional assay, we monitored the transient transactivation of a reporter plasmid (encoding the luciferase gene under the control of the thymidine kinase promoter with an ERE) upon cotransfection with different ER variants. Moreover, a translocation test was designed to "look" directly at the ER protein: we took advantage of an ER mutant (C447 S) (Neff et al. 1994), which displays ligand-dependent nuclear localization. We were able to show activation of the human or mouse ER by a serie of agents including okadaic acid, cAMP, and EGF. We then characterized the signaling pathway by which EGF activates ER. Using trans-dominant negative mutants of Ras (N17) and mitogen-activated protein kinase kinase (MAPKK), activation of the luciferase reporter gene or translocation of the ER variant C447 S to the nucleus upon addition of EGF could be blocked. Conversely, activation of hER could be mimicked by a constitutively active mutant of MAPKK. Thus, ligand-independent activation of hER by EGF works through the MAP kinase cascade (Bunone et al. 1996). The domain of hER involved in ligand-independent activation by EGF was mapped using chimeras between GR and ER, and ER mutants. We showed that the N-terminal A/B domain of ER containing AF-1 is the target region for EGF signaling. Within this domain, serine 118 plays a key role in this process. The mutant S118 A (the serine 118 has been replaced by an alanine) lacks AF-1 activity (Ali et al. 1993) and is insensitive to EGF (Bunone et al. 1996). Serine 118 is a major phosphorylation site in vitro and in vivo (Ali et al. 1993; Le Goff et al. 1994) and is a target site for MAP kinase in vitro (Kato et al. 1995). Thus, the ligand-independent activation of ER by EGF is likely to involve a direct phosphorylation at serine 118 by MAP kinase. While this phosphorylation is not sufficient for ER activation, other growth factors may affect ER activity by targeting the same serine.

6.3 Activation of the Estrogen Receptor by Dopamine

Both hER and cPR are also activated in the absence of cognate ligands by dopamine or agents that activate protein kinase A, such as analogues of cAMP (dibutyryl cAMP or 8-Bromo-cAMP) (Power et al. 1991; Aronica and Katzenellenbogen 1993; Smith et al. 1993; Bunone et al. 1996). Dopamine action is mediated by its interaction with membrane-bound receptors coupled to G proteins. Several dopamine receptors have been identified: they belong to two main receptor subtypes, D1 and D2, which stimulate and inhibit adenylate cyclase, respectively (Kabebian and Calne 1979; Jackson and Westlind-Danielsson 1994). For cPR it was reported that ligand-independent activation may occur through stimulation of the D1 receptor subtype, based on the observation that a selective D1 receptor agonist (SKF-38393) stimulates cPR while a selective D2 receptor agonist (quinpirole) does not (Power et al. 1991). Interestingly, activation of cPR by dopamine cannot be accounted for by the activation of adenylate cyclase and the cAMP increase alone since addition of isoprotenerol, which is coupled to adenylate cyclase through a β-adrenergic receptor and also leads to an increase in cAMP level, does not activate cPR in a ligand-independent manner (Power et al. 1991). This suggests a more complex signaling mechanism for dopamine than just via an elevation of intracellular cAMP.

6.3.1 Mapping of the Dopamine Target Domain

Our interest is, firstly, to map the domain(s) and the phosphorylation sites in hER that are targeted by dopamine signaling and, secondly, to dissect the dopamine signaling pathway involved in ER activation. We have started to characterize ER activation by the selective dopamine D1 receptor agonist SKF-38393 in SK-Br-3 breast cancer cells using the transcriptional assay (see Sect. 6.2). SK-Br-3 cells were chosen as tester cells because of their strong response to dopamine (Fig. 1). Different ER variants have been used in this study (Fig. 2). We first reproduced

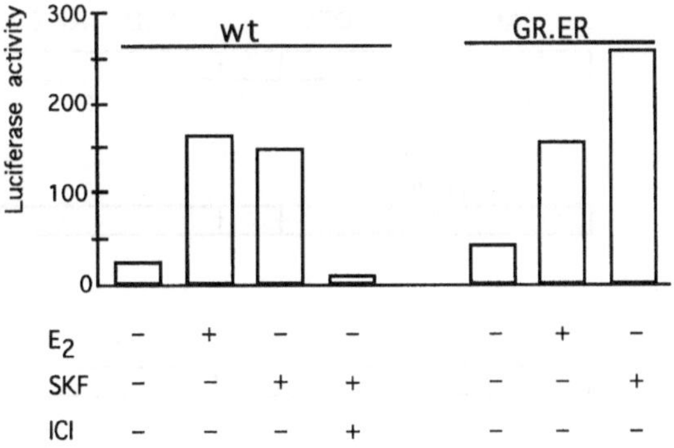

Fig. 1. The hormone-binding domain of the human estrogen receptor is necessary for activation by dopamine. Human breast cancer SK-Br-3 cells were transfected with plasmids encoding the glucocorticoid receptor N-terminal domain (*GR.ER*) or wild-type estrogen receptor (*wt*), and the firefly luciferase encoding reporter plasmid (EREtkluc) (Bunone et al. 1996). Before transfection, the cells were maintained for 48 h in DMEM without phenol red supplemented with 10% charcoal-treated fetal calf serum. Plates were transfected using the calcium phosphate coprecipitation technique and processed as described (Bunone et al. 1996). The transfections were standardized using the *Renilla* luciferase encoded by plasmid pRL/CMV (Promega, Madison, WI). E_2, 100 nM 17β-estradiol; SKF, 100 μM SKF-38393 (specific agonist for dopamine D1 receptor, RBI); ICI, 1 μM ICI164'384 (ICI Pharmaceuticals)

the observation that ER can be activated by the dopamine agonist SKF-38393 (Fig. 1). This activation is specifically inhibited by the pure antiestrogen ICI 164'384.

6.3.1.1 The ER HBD Is Required for Dopamine Activation

Earlier studies on PR indicated that the dopamine signal may be targeted to the HBD of the PR since serine 628 was crucial for activation (Power et al. 1991). To map the dopamine responsive domain in ER, we used chimeras between the responsive hER and the unresponsive GR (see Fig. 2) (Power et al. 1991). A chimeric receptor with the ER HBD and the GR N-terminal domain (GR.ER) also responds to the dopamine agonist SKF-38393, while the EGF-inducible chimera with the GR HBD and the hER N-terminus (Bunone et al. 1996) is only poorly activated by the dopamine agonist (data not shown). These data suggest

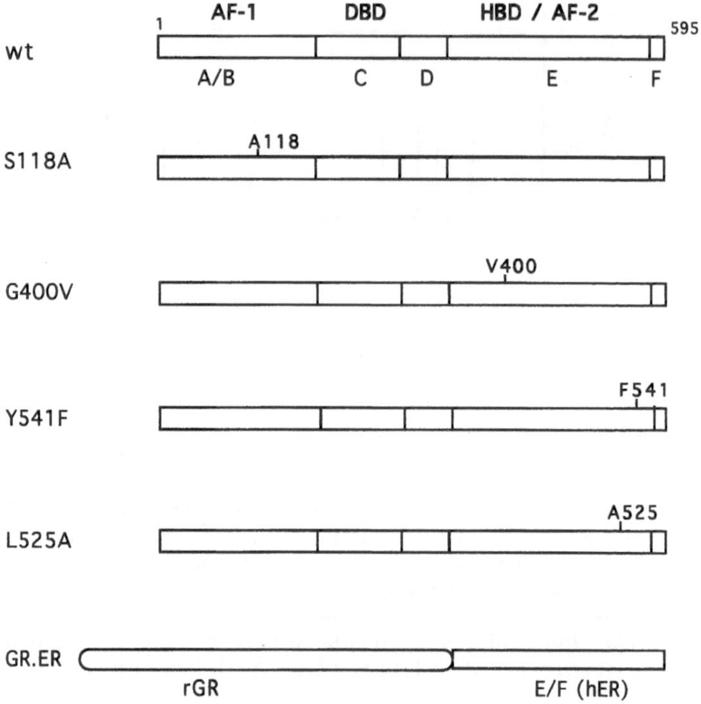

Fig. 2. Structure of estrogen receptor (ER) derivatives and chimera used in this study. The domains of ER (*A/B*, *C–F*), as well as the transcriptional transactivation functions (*AF-1* and *AF-2*), the DNA-binding domain (*DBD*), and the hormone binding domain (*HBD*) are shown. Amino acid substitutions are indicated above the *open boxes*. Wild-type (*wt*), S118 A, G400 V, Y541F, and L525 A are encoded by plasmid HEG0 (Tora et al. 1989), HE457 (Ali et al. 1993), HE0 (Greene et al. 1986), pMORY541F (White et al. 1997), and pCMVhERL525 A (Ekena et al. 1996), respectively. In chimera the glucocorticoid receptor N-terminal domain (*GR.ER*) chimera, the HBD of rat GR (*rGR*) has been replaced by the equivalent domain of ER (Bunone et al. 1996). Except for Y541F (which is the mouse ER), ER mutants used in this study are derived from the human ER

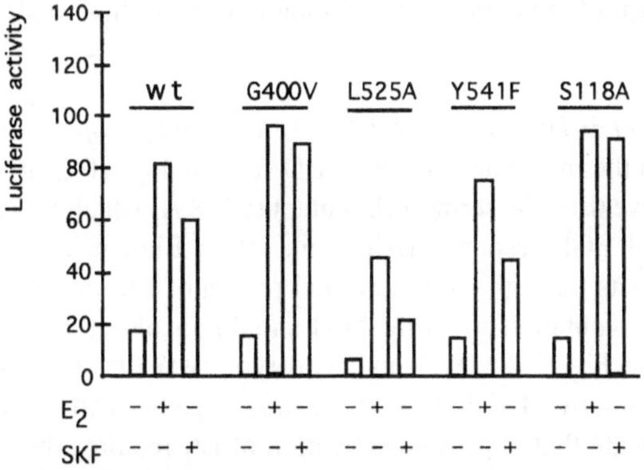

Fig. 3. Activation of different human estrogen receptor variants by dopamine. Transactivation assays were performed in SK-Br-3 cells. Before transfection, the cells were maintained for 48 h in DMEM without phenol red supplemented with10% charcoal-treated fetal calf serum. Plates were transfected using the calcium phosphate coprecipitation technique and processed as described (Bunone et al. 1996). The transfections were standardized using the *Renilla* luciferase encoded by plasmid pRL/CMV (Promega, Madison, WI). E_2, 100 nM 17β-estradiol; SKF, 100 μM SKF-38393 (specific agonist for dopamine D1 receptor, RBI). For L525 A, 1 μM E2 was used. *wt*, wild-type

that the hER HBD may be sufficient for dopamine activation, in parallel with the results for PR. Since the dopamine responsive C-terminal domain of ER overlaps with a functional ligand-binding domain, we could not formally rule out that dopamine exerts its effect on receptors having the ligand-binding site occupied. However, the fact that the activation of the GR.ER chimera by SKF-38393 is seen in cells that have been rigorously depleted of estrogen makes this unlikely. To exclude the possibility that low levels of estrogen present in the medium synergize with dopamine for ER activation, we used a mutant of ER carrying a point mutation in the HBD (L525 A) (Figs. 2 and 3). This mutant is only activated by very high concentrations of estrogens (Ekena et al. 1996), for example 1 μM estradiol as shown in Fig. 3. Addition of SKF-38393 also activated the ER mutant, suggesting strongly that dopamine activation does not require residual estrogen and is not a form of synergy whereby dopamine amplifies weak activation by steroid.

6.3.1.2 AF-1 Is Dispensable for ER Activation by Dopamine

To confirm the importance of the HBD (containing AF-2) in dopamine signaling, we used the serine 118 mutant (S118 A) which lacks the AF-1 activity and EGF response (Ali et al. 1993; Bunone et al. 1996). If dopamine acts through the C-terminal region of ER, and thus possibly AF-2, then a mutation affecting AF-1 should not alter this signaling. As presented in Fig. 3, S118 A and wild-type (wt) ER showed identical transactivation by SKF-38393. These data, together with those given in Fig. 1, indicate that dopamine activation of ER requires the C-terminal HBD and possibly AF-2 (the requirement for AF-2 needs to be confirmed with an AF-2 mutant).

6.3.1.3 Towards Characterizing Key Residues in the HBD

To begin mapping key residues in the HBD involved in dopamine activation, we looked for ER mutants that might be unresponsive to SKF-38393. Tyrosine 541 in the mouse ER (corresponding to Y537 in hER) seemed to be an interesting candidate. When mutated to alanine, serine, glutamic acid, or aspartic acid, this residue confers estrogen-independent constitutive activity to the receptor (Weis et al. 1996; White et al. 1997). Because of the location of this residue at the N-terminus of a conserved helix, which forms a major part of the ligand-dependent activation surface in the ER, it has been postulated that it is required to maintain the receptor in a transcriptionally inactive state in the absence of the hormone (White et al. 1997). We used the mouse ER carrying a phenylalanine for tyrosine substitution (Y541F): Unlike other Y541 mutants, this mutant is not constitutively activated and can thus be used in our transactivation assay. As presented in Fig. 3, the mutant Y541F can be activated by estradiol as efficiently as wt, and addition of the dopamine agonist SKF-38393 to the media allowed activation of both wt and mutant in the absence of E_2. Thus, dopamine does not require a tyrosine at residue 541 for ER activation. Another potentially interesting residue is glycine 400 which has been claimed to be unresponsive to dopamine when mutated to valine (Smith et al. 1993). We therefore tested this mutant (called G400 V) in SK-Br-3 cells. Surprisingly, unlike what has been reported by Smith et al. (1993), G400 V can be activated as efficiently as wt by E_2 and SKF-38393. This discrepancy may be due to the cell line used (SK-Br-3 in this study versus Hela cells).

6.4 Conclusions

Here we report that ER can be activated in the absence of steroid by SKF-38393 (a selective agonist for dopamine D1 receptor) and that the HBD of the receptor is required. Interestingly, the data presented in this study show that the activation of the ER by dopamine and EGF requires different domains (El-Tanani and Green 1997). This observation probably reflects differences in the signaling cascades. For EGF, the MAP kinase cascade is activated and MAP kinase phosphorylates the serine 118 in the N-terminal domain of the ER (Kato et al. 1995; Bunone et al. 1996). For dopamine, the signaling pathway is less well defined. The dopamine receptor of the D1 subclass is coupled to G proteins which activates adenylate cyclase, leading to an increase of intracellular cAMP concentration (Kabebian and Calne 1979; Jackson and Westlind-Danielsson 1994) and to the activation of protein kinase A. The following steps in this pathway are still uncharacterized, but in the case of ER they may most probably target the HBD and therefore may not involve MAP kinases.

This study confirms that ER, as well as other nuclear receptors, can be activated by a variety of alternative pathways from membrane receptors. Under most physiological conditions, steroid receptors may be activated by both ligand-dependent and -independent pathways, resulting in synergistic effect at target genes. But under some conditions, either physiological or pathological, steroid receptors may be mainly regulated by signaling from membrane receptors triggering multiple downstream kinases and/or phosphatases. This "cross-talk" has implications for breast (Picard et al. 1997) and endometrial cancers and cannot be ignored if new therapies are to be designed for the future.

Acknowledgments. This work was supported by the Swiss National Science Foundation, the foundation "Recherche suisse contre le cancer", and the Canton de Genéve.

References

Ali S, Metzger D, Bornert JM, Chambon P (1993) Modulation of transcriptional activation by ligand-dependent phosphorylation of the human oestrogen receptor A/B region. EMBO J 12:1153–1160

Aronica SM, Katzenellenbogen BS (1993) Stimulation of estrogen receptor-mediated transcription and alteration in the phosphorylation state of the rat uterine estrogen receptor by estrogen, cyclic adenosine monophosphate, and insulin-like growth factor-I. Mol Endocrinol 7:743–752

Bagchi MK, Tsai S-Y, Tsai M-J, O'Malley BW (1992) Ligand and DNA-dependent phosphorylation of human progesterone receptor in vitro. Proc Natl Acad Sci USA 89:2664–2668

Bunone G, Briand P-A, Miksicek RJ, Picard D (1996) Activation of the unliganded estrogen receptor by EGF involves the MAP kinase pathway and direct phosphorylation. EMBO J 15:2174–2183

Denner LA, Weigel NL, Maxwell BL, Schrader WT, O'Malley BW (1990) Regulation of progesterone receptor-mediated transcription by phosphorylation. Science 250:1740–1743

Ekena K, Weis KE, Katzenellenbogen JA, Katzenellenbogen BS (1996) Identification of amino acids in the hormone binding domain of the human estrogen receptor important in estrogen binding. J Biol Chem 271:20053–20059

El-Tanani MK, Green CD (1997) Two separate mechanisms for ligand-independent activation of the estrogen receptor. Mol Endocrinol 11:928–937

Evans RM (1988) The steroid and thyroid hormone receptor superfamily. Science 240:889–895

Greene GL, Gilna P, Waterfield M, Baker A, Hort Y, Shine J (1986) Sequence and expression of human estrogen receptor complementary DNA. Science 231:1150–1154

Ignar-Trowbridge DM, Teng CT, Ross KA, Parker MG, Korach KS, McLachlan JA (1993) Peptide growth factors elicit estrogen receptor-dependent transcriptional activation of an estrogen-responsive element. Mol Endocrinol 7:992–998

Ignar-Trowbridge DM, Pimentel M, Teng CT, Korach KS, McLachlan JA (1995) Cross talk between peptide growth factor and estrogen receptor signaling systems. Environ Health Perspect 103:35–38

Ignar-Trowbridge DM, Pimentel M, Parker MG, McLachlan JA, Korach KS (1996) Peptide growth factor cross-talk with the estrogen receptor requires the A/B domain and occurs independently of protein kinase C or estradiol. Endocrinology 137:1735–1744

Jackson DM, Westlind-Danielsson A (1994) Dopamine receptors: molecular biology, biochemistry and behavioural aspects. Pharmacol Ther 64:291–370

Kabebian JW, Calne DB (1979) multiple receptors for dopamine. Nature 277:93–96

Kato S, Endoh H, Masuhiro Y, Kitamoto T, Uchiyama S, Sasaki H, Masushige S, Gotoh Y, Nishida E, Kawashima H, Metzger D, Chambon P (1995) Activation of the estrogen receptor through phosphorylation by mitogen-activated protein kinase. Science 270:1491–1494

Kumar V, Green S, Stack G, Berry M, Jin J-R, Chambon P (1987) Functional domains of the human estrogen receptor. Cell 51:941–951

Le Goff P, Montano MR, Schodin DJ, Katzenellenbogen BS (1994) Phosphorylation of the human estrogen receptor – identification of hormone-regulated sites and examination of their influence on transcriptional activity. J Biol Chem 269:4458–4466

Ma ZQ, Santagati S, Patrone C, Pollio G, Vegeto E, Maggi A (1994) Insulin-like growth factors activate estrogen receptor to control the growth and differentiation of the human neuroblastoma cell line SK-Er3. Mol Endocrinol 8:910–918

Miner JN, Yamamoto KR (1991) Regulatory crosstalk at composite response elements. Trends Biochem Sci 16:423–426

Neff S, Sadowski C, Miksicek RJ (1994) Mutational analysis of cysteine residues within the hormone-binding domain of the human estrogen receptor identifies mutants that are defective in both DNA-binding and subcellular distribution. Mol Endocrinol 8:1215–1223

Newton CJ, Buric R, Trapp T, Brockmeier S, Pagotto U, Stalla GK (1994) The unliganded estrogen receptor (ER) transduces growth factor signals. J Steroid Biochem Mol Biol 48:481–486

Patrone C, Ma ZQ, Pollio G, Agrati P, Parker MG, Maggi A (1996) Cross-coupling between insulin and estrogen receptor in human neuroblastoma cells. Mol Endocrinol 10:499–507

Picard D, Bunone G, Liu JW, Donzé O (1997) Steroid-independent activation of steroid receptors in mammalian and yeast cells and in breast cancer. Biochem Soc Trans 25:597–602

Pietras RJ, Arboleda J, Reese DM, Wongvipat N, Pegram MD, Ramos L, Gorman CM, Parker MG, Sliwkowski MX, Slamon DJ (1995) HER-2 tyrosine kinase pathway targets estrogen receptor and promotes hormone-independent growth in human breast cancer cells. Oncogene 10:2435–2446

Power RF, Mani SK, Codina J, Conneely OM, O'Malley BW (1991) Dopaminergic and ligand-independent activation of steroid hormone receptors. Science 254:1636–1639

Saatcioglu F, Claret FX, Karin M (1994) Negative transcriptional regulation by nuclear receptors. Semin Cancer Biol 5:347–359

Smith CL, Conneely OM, O'Malley BW (1993) Modulation of the ligand-independent activation of the human estrogen receptor by hormone and antihormone. Proc Natl Acad Sci USA 90:6120–6124

Takimoto GS, Tasset DM, Eppert AC, Horwitz KB (1992) Hormone-induced progesterone receptor phosphorylation consists of sequential DNA-independent and DNA-dependent stages: analysis with zinc finger mutants and the progesterone antagonist ZK98299. Proc Natl Acad Sci USA 89:3050–3054

Tora L, Mullick A, Metzger D, Ponglikitmongkol M, Park I, Chambon P (1989) The cloned human estrogen receptor contains a mutation which alters its hormone binding properties. EMBO J 8:1981–1986

Tsai M-J, O'Malley BW (1994) Molecular mechanisms of action of steroid/thyroid receptor superfamily members. Annu Rev Biochem 63:451–486

Webster NJG, Green S, Jin JR, Chambon P (1988) The hormone-binding domains of the estrogen and glucocorticoid receptors contain an inducible transcription activation function. Cell 54:199–207

Weigel NL, Zhang Y (1997) Ligand-independent activation of steroid hormone receptors. J Mol Med (in press)

Weis KE, Ekena K, Thomas JA, Lazennec G, Katzenellenbogen BS (1996) Constitutively active human estrogen receptors containing amino acid substitutions for tyrosine 537 in the receptor protein. Mol Endocrinol 10:1388–1398

White R, Sjoberg M, Kalkhoven E, Parker MG (1997) Ligand-independent activation of the oestrogen receptor by mutation of a conserved tyrosine. EMBO J 16:1427–1435

7 The ER/AP1 Pathway: A Window on the Cell-Specific Estrogen-like Effects of Antiestrogens

P. Webb, M.-R. Keneally, J. Shinsako, R. Uht, C. Anderson, K. Paech, T.S. Scanlan, and P.J. Kushner

7.1 Introduction

We study an unusual pathway of estrogen receptor (ER) action, in which ER indirectly alters the transactivation capacity of the jun/fos complex (AP1 proteins), rather than stimulating gene expression by directly binding DNA. Here, we review the rationale for studying this pathway and the reasons for thinking that elucidation of ER/AP1 interactions will help us design the next generation of antiestrogens and hormone replacement therapies.

7.2 Estrogen-like Behavior of Antiestrogens

Estrogens stimulate cell growth in the breast and uterus (Jordan 1992; Morrow and Jordan 1993). They are also known to stimulate growth of early stage tumors that arise in both organs. Estrogen signals are transduced by two estrogen receptors (ERα and ERβ) which are both conditional transcription factors. It is believed that estrogens trigger growth responses by stimulating ER activity which, in turn, alters the cellular pattern of gene expression. The estrogen induced gene products then lead to enhanced growth (Gronemeyer et al. 1992).

Although the genes involved in estrogen regulated growth are still under investigation, the fact that a transcription factor (ER) lies upstream of complex estrogen induced growth pathways makes the ER molecule an attractive target for intervention. Antiestrogens, such as tamoxifen, that compete for the estrogen binding pocket of ER, inhibit breast cancer growth and have been successfully used as breast cancer therapeutics (Green 1990; Gronemeyer et al. 1992; Katzenellenbogen et al. 1996). Tamoxifen also reduces the incidence of breast cancer, and it is presently proposed that tamoxifen be used to prevent tumors in women with high cancer risk (Anon. 1992).

The use of tamoxifen as a preventative agent, however, is confounded by the fact that it also behaves like estrogen in some conditions (Morrow and Jordan 1993). Tamoxifen stimulates cell growth and gene expression in the uterus, and has been linked to an increased risk of uterine cancer (Kedar et al. 1994). Even more disturbingly, tamoxifen may switch from inhibition to stimulation of breast cancer growth during tumor progression (Horwitz 1994). On the other hand, tamoxifen

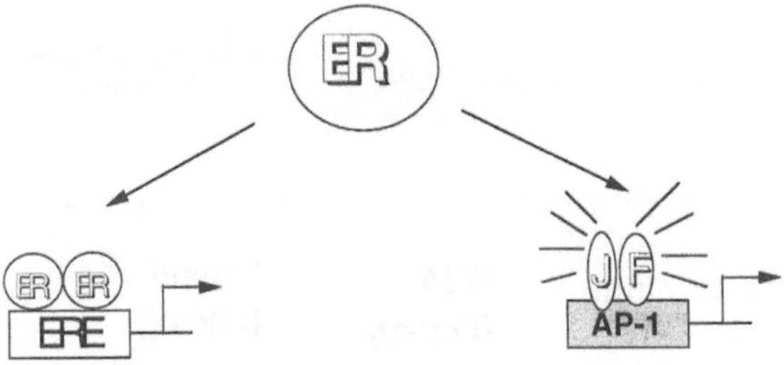

Fig. 1. Two mechanisms of estrogen response. The estrogen receptor (*ER*) can activate transcription by binding to estrogen response elements (*ERE*) in estrogen-responsive promoters or by altering the activity of heterologous transcription factors at alternate response elements, including AP1 or jun/fos (*J*/*F*)

also shows beneficial estrogen like effects, such as reversal of bone loss and reduced risk of heart disease, that are compatible with hormone replacement therapy.

Our long term aim is to identify improved antiestrogens, that lack estrogen-like effects in uterus and late stage breast tumors. We would also like to identify new hormone replacement therapies, which would preserve the beneficial effects of estrogens, while lacking the possible cancer risk of estrogen supplements.

7.3 Models of Tamoxifen Agonism

Two main mechanisms of estrogen response are known. In the classical pathway estrogen response (Fig. 1), the estrogen liganded ER binds a cognate estrogen response element (ERE) that marks estrogen responsive genes (Parker 1990). From this location, the ER is thought to recruit coactivators that mediate transcription from nearby promoters. ER also stimulates gene expression through alternate response elements, such as AP1, via poorly defined protein–protein interactions (see below).

Tamoxifen action in the classical pathway is well understood (Green 1990, Gronemeyer et al. 1992; Katzenellenbogen et al. 1996). Tamoxifen allows the receptor to bind DNA, but interferes with the ability of ER to activate genes. Like most members of the steroid receptor

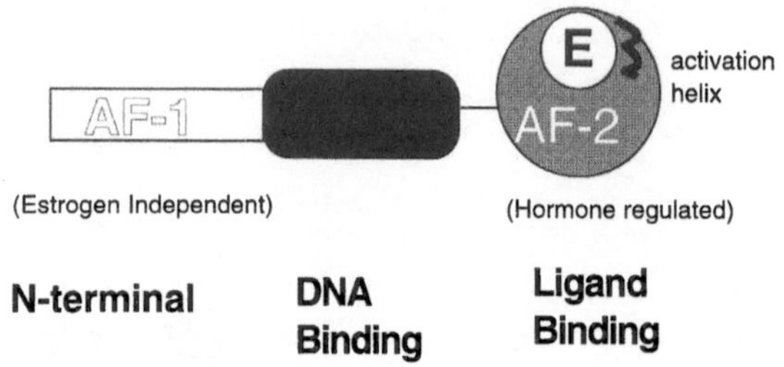

Fig. 2. Structural organization of the estrogen receptor (ER). The ER consists of three separable domains, the AB domain, the DNA-binding domain, and the ligand-binding domain (LBD). Tamoxifen blocks estrogen (*E*) binding to the LBD, but allows the AB domain transactivation function AF1 to work at classical estrogen response elements

superfamily, ER possesses a tripartite structure composed of three domains, the AB domain, the DNA binding domain (DBD), and the ligand binding domain (LBD) (Fig. 2). ER dependent transcriptional activation requires two separable activation functions (AF1 and AF2). Estrogen binding allows the ER-LBD to display AF2 in a manner that allows binding of p160 coactivator proteins such as SRC-1 (Onate et al. 1995) GRIP1/TIF2 (Hong et al. 1996; Voegel et al. 1996) and pCIP/RAC3/ACTR/AIB1 (Torchia et al. 1997; Li et al. 1997; Chen et al. 1997; Anzick et al. 1997). Tamoxifen and other antiestrogens do not allow the ER-LBD to bind these coactivators and, consequently, do not allow ER to activate transcription.

Tamoxifen agonism in the context of the classical pathway arises from AF1 (Berry et al. 1990). Tamoxifen allows ER to bind DNA, and consequently brings AF1 to estrogen regulated promoters. Usually AF1 is weak and the tamoxifen-liganded ER does not activate genes. AF1, however, can be strongly active in three situations. It shows strong cell type specific activity, strong promoter specific activity and, because the AB domain contains a cluster of MAP kinase phosphorylated serines, under conditions of MAP kinase stimulation (Ali et al. 1993). Consequently, the tamoxifen liganded ER shows strong activity in each of these three conditions.

The recent cloning of a second ER gene (ERβ) introduced another possible source of antiestrogen agonist effects (Katzenellenbogen and Korach 1997). The ERα and ERβ genes show little conservation in their AB domains, and initial studies have shown that tamoxifen behaves as a pure antagonist of ERβ function (Tremblay et al. 1997). Nonetheless, ERβ does possess a site in its amino terminus that is regulated by MAP kinases, suggesting that ERβ AF1 may also be subject to cell surface regulation (Tremblay et al. 1997).

7.4 Our Goal: A Paradigm for Tamoxifen Activity

Understanding how estrogen responsive growth genes are regulated will help us to design more specific ways to interfere with estrogen action. To begin to address this question we decided to ask whether a simple gene expression paradigm would parallel the effects of estrogen and tamoxifen on cell growth. Such a paradigm could help us understand the estrogen-like effects of tamoxifen without necessarily identifying every gene downstream of ER action. We were interested in the possibility that AF1 activity at classical EREs would correlate with tamoxifen effects on growth. Because antiestrogen action at alternate response elements was not known at the time we began our study, we were also interested in finding out how antiestrogens affected the AP1 pathway.

7.5 Tamoxifen Action in Steroid Sensitive Tumor Cells: Usefulness of the AP1 Paradigm

We began by testing how estrogen and tamoxifen affected gene expression in two endocrine-sensitive tumor cell lines (Webb et al. 1995). To model estrogen responsive growth we used MCF-7 breast cells, which grow in tissue culture and nude mice in response to estrogen, and whose growth is antagonized by tamoxifen. To model tamoxifen responsive growth we used Ishikawa uterine tumor cells which grow in tissue culture and nude mice in response to either estrogen or tamoxifen.

Initially, we tested a standard reporter gene with a classical ERE linked upstream of the herpes simplex virus-TK promoter (ERE-TK-CAT). A typical result is shown in Fig. 3. ERE-TK-CAT was predomi-

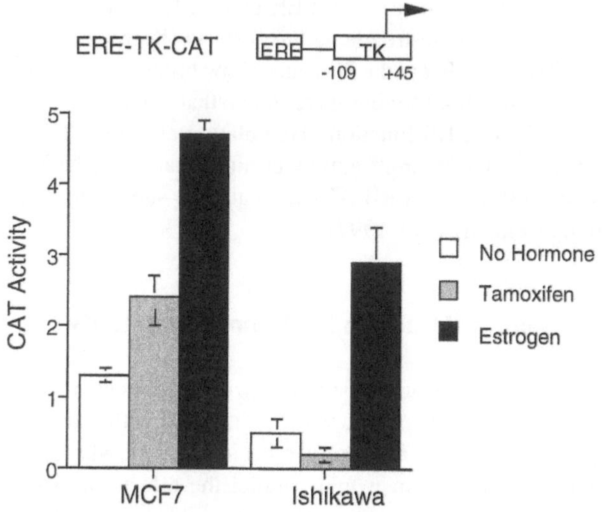

Fig. 3. Estrogen receptor action at classical estrogen response elements (*ERE*) in endocrine-sensitive tumor cell lines. A reporter gene vit A2 ERE-TK-CAT was transfected into MCF7 cells and Ishikawa cells. CAT activity was determined in extracts of cells treated with estrogen (10 n*M*) or tamoxifen (5 μ*M*) after 18 h. The reporter gene responded predominantly to estrogen in both cell lines

nantly estrogen responsive in both MCF7 and Ishikawa cells. Surprisingly, tamoxifen showed some activity in MCF7 (discussed in Sect. 7.6) and none in Ishikawa. Similar results were obtained with several different Ishikawa cell sublines, several different reporter genes, a range of serum conditions, and in the presence of epidermal growth factor and phorbol esters, which stimulate MAP kinase pathways. Thus, contrary to our expectations, Ishikawa cells did not show strong generalized AF1 activity.

We next tested whether ER regulation of an AP1-responsive reporter would correlate with a tamoxifen-dependent growth response. Figure 4 shows that AP1-dependent reporters were estrogen responsive in MCF7 cells. Similar results were previously obtained at Henri Rochefort's laboratory (Philips et al. 1993). In Ishikawa cells, however, AP1-respon-

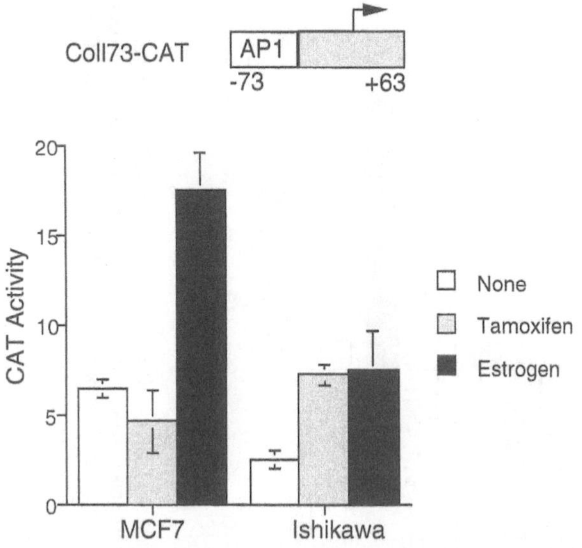

Fig. 4. Estrogen receptor action at AP1 sites in endocrine-sensitive tumor cell lines. AP1 responsive reporters were transfected into MCF7 cells and Ishikawa cells. AP1 responsive reporters responded mostly to estrogen in MCF7 cells and both estrogen and tamoxifen in Ishikawa cells

sive genes responded to estrogen and tamoxifen. Thus, estrogen and tamoxifen effects on growth correlated with AP1-responsive transcription.

We went on to find large estrogen and antiestrogen effects at AP1 sites in many different cell types, but never in breast cells. ER action at AP1-responsive reporters has also been demonstrated by several other groups in studies that pre- and post-dated our first publication (Gaub et al. 1990; Philips et al. 1993; Umayahara 1994; Uht et al. 1997). ER action at AP1 sites may therefore be an important pathway of ER action in nature.

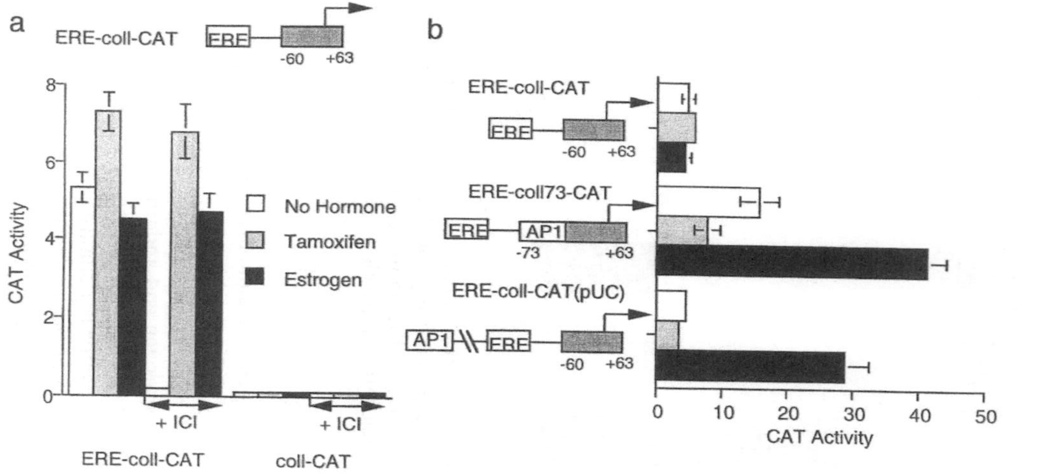

Fig. 5a,b. Strong AF1 activity in breast cells. **a** Activity of ERE-coll-CAT in MCF7 cells. The estrogen response element (*ERE*) conferred strong tamoxifen responses. **b** AP1 sites and EREs in *cis*-restored estrogen response in MCF7 cells. Either the collagenase AP1 site, or the backbone pUC AP1 site gave a similar effect

7.6 Strong AF1 Activity in Breast Cells

Tamoxifen activity at classical EREs was strongest in MCF7 cells (Fig. 3). This was illustrated even better by a reporter gene in which we replaced the collagenase AP1 site with a classical ERE (ERE-coll-CAT; Fig. 5a). The ERE conferred constitutive activity upon the collagenase core promoter that was unaffected by either tamoxifen or estrogen. A pure antiestrogen ICI 164,384 (ICI), however, abolished constitutive activation, and tamoxifen and estrogen both reversed this effect. Several alternate reporter genes, with EREs upstream of minimal promoters, behaved similarly (data not shown). Thus, ER action upon simple ERE responsive reporters in MCF7 cells was marked by strong AF1 activity, even though tamoxifen blocked cell growth under the same conditions. This suggests that strategies to block AF1 activity need not always interfere with tamoxifen effects on cell growth, unless the AF1 regulated gene is essential for growth response.

We were able to restore estrogen response in MCF7 cells by placing the ERE and AP1 site in the *cis* position (Fig. 5b). It is known that pUC vectors contain an AP1 site, which we removed from all of our reporter genes. We found ERE-coll-CAT permitted a strong estrogen response when the backbone plasmid was unmodified. Perhaps this AP1 site contributed to previous demonstrations of estrogen responses in MCF7 cells which used pUC-based reporter genes. This further underlines the importance of studying ER action at AP1 sites.

7.7 Predictive Value of the AP1 Paradigm

Because ER action at AP1 sites correlated with ER effects on growth, we suggested that studying ER action at AP1 sites would be a useful predictor of antiestrogen behavior. We further examined this idea by asking how other antiestrogens behaved at AP1 sites.

7.7.1 Nonclassical Antiestrogenicity: Other Classes of Steroids

Estrogen action is often opposed by other classes of steroids in vivo. For example, glucocorticoids, progestins and androgens show nonclassical antiestrogenicity in breast tumors and the uterus (Zhou et al. 1989). While ER activates AP1, other steroid receptors repress AP1. Could opposing steroid influences be mediated at the level of transcription through AP1 sites? Rosalie Uht examined this question using combinations of ERα and glucocorticoid receptor (GR) or progesterone receptor (PR) (Uht et al. 1997). She found that estrogen reduced glucocorticoid inhibition, and glucocorticoids blunted estrogen stimulation. The relative amounts of ER and GR/PR determined the direction of the AP1 response. A similar effect was documented for ER and PR. Thus, ER and GR/PR opposed each other's effects at AP1, recapitulating the nonclassical antiestrogenicity of glucocorticoids and progestins.

7.7.2 Alternate Antiestrogens

Raloxifene. We were intrigued by reports that a type II antiestrogen called raloxifene, which is related to tamoxifen, showed less uterotropic activity (Yang et al. 1996). With our colleagues in the Scanlan lab, we tested the effects of this compound at AP1 sites (Fig. 6; Paech et al. 1997). Raloxifene showed little activity (about 10%–20% of tamoxifen) at an AP1 responsive reporter gene in uterine cells. Thus, raloxifene uterotropic activity was a useful reverse predictor of the activity of the raloxifene-liganded alpha receptor at AP1 sites, further cementing the correlation between ER action at AP1 and ER growth effects.

"Pure" Antiestrogens. ICI 164,384 and 182,782 (ICI), are "pure" antiestrogens (Wakeling 1993). They both consist of an estradiol moiety with a long aliphatic chain that is likely to project from the ER molecule. While the ICI drugs completely abolish ER activity at EREs, they show agonist activity at AP1 sites (Fig. 6). So far, the ICI drugs behave as pure antiestrogens in vivo. Does this discrepancy mean that the AP1 paradigm is not active in nature?

We think that the key to this question lies in determining why ICI acts as a pure antiestrogen. ICI blocks estrogen action at classical EREs by inhibiting AF2 interaction with coactivators (Cavaillès et al. 1994) and residual AF1 activity by an unknown mechanism (McDonnell et al. 1995). ICI, however, also increases ER protein turnover and reduces ER

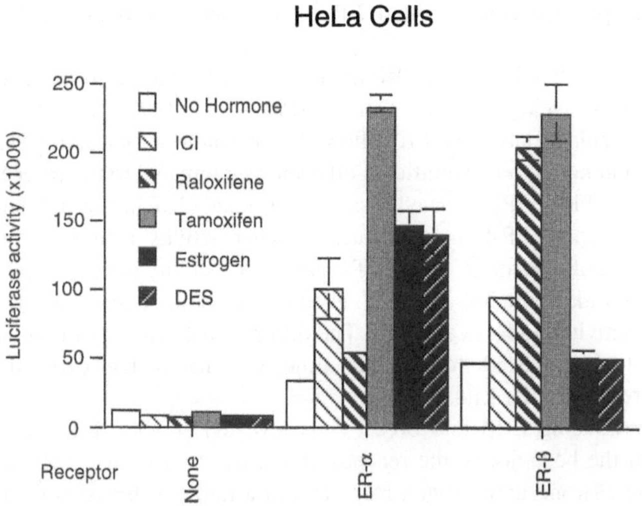

Fig. 6. Effect of ER ligands and isoforms at AP1 sites. HeLa cells were transfected with empty vector, or vectors for ERα or rat ERβ. Cells were treated with vehicle, ICI, raloxifene, tamoxifen, estradiol, or diethylstilbestrol (*DES*)

titre (Dauvois et al. 1992). It is plausible that ICI acts as a pure antiestrogen in vivo primarily because it eliminates ER protein. Indeed, the literature now contains many examples of ICI-induced genes, such as PR (Jamil et al. 1991), TGFβ3 (Yang et al. 1996) and quinone reductase (Montano and Katzenellenbogen 1997). We predict ICI will show agonist activity in cases where ER effects are mediated by alternate response elements and where it fails to completely eliminate ER protein.

7.8 ERβ and the AP1 Pathway

Recently, a second ER (ERβ) gene was isolated (Kuiper et al. 1996; Mosselman et al. 1996; Tremblay et al. 1997). The predicted protein strongly resembled ERα in the DBD and LBD regions. Accordingly, ERβ recognized an ERE, activated transcription from classical estrogen-responsive genes, bound ER ligands and heterodimerized with ERα.

We wondered whether ERβ might also mediate antiestrogen agonist effects. In an elegant series of studies (Paech et al. 1997), Kolja Paech in Tom Scanlan's lab found ERβ allowed strong antiestrogen action at AP1 sites and noted three significant differences compared to the α receptor (Fig. 6). First, ERβ showed less activity than ERα in the unliganded state. Second, ERβ showed much stronger activity than ERα in the presence of raloxifene. Third, ERβ gave little activity in the presence of agonists like estradiol and DES, even though both compounds elicited ERβ activity at classical EREs. The different behaviors of the α and β receptors suggested, for the first time, why nature had evolved two apparently similar ERs.

Introducing ERβ into breast cells revealed another difference between the behavior of the receptor isoforms at AP1 sites. ERα never allowed strong antiestrogen responses in a range of breast cells, even when transiently transfected at high levels. In contrast, even low levels of ERβ gave strong antiestrogen responses in breast cells, especially with raloxifene.

Together, the findings suggest intriguing possibilities for ERβ function. ERβ is likely to dampen AP1-mediated responses in the presence of estradiol and may play in a role in some instances of observed antiestrogen agonism. For example, ERβ is expressed in bone (Onoe et al. 1997) and in some breast tumors (Dotzlaw et al. 1997), suggesting it could play a role in raloxifene agonist effects in bone and in acquired antiestrogen resistance in breast. We also expect that the ERβ story will yield yet another layer of complexity. Recent observations have suggested that longer forms of the ERβ protein may exist in nature. We await the cloning and characterization of these forms with interest. Understanding how these forms of ERβ are regulated and how they activate AP1 will undoubtedly provide insights into their function.

7.9 Does the ER/AP1 Pathway Really Regulate Growth Genes?

Understanding ER action at AP1 sites will help us understand how antiestrogens show strong agonist effects. Could we take the argument a stage further and suggest that ER growth effects are actually mediated predominantly by AP1 responsive genes? In principle, this idea is attractive because AP1 is known to transduce growth signals from the cell surface. ER action at AP1 would allow subversion of pathways that would already be sensitive to other growth stimulatory signals. Here, we review examples of estrogen-regulated genes that support the idea that AP1 sites are important in estrogen-regulated growth.

7.9.1 Insulin-like Growth Factor I

Insulin-like Growth Factor I (IGF-I) production is stimulated by either estrogen or tamoxifen in the uterus (Huynh and Pollak 1993). The IGF-I promoter was found to be estrogen responsive, lack classical EREs, and contain a consensus AP1 site which mediates ER effects (Umayahara et al. 1994). Thus, IGF-I fulfills our predictions about the nature of estrogen-responsive growth regulatory genes.

7.9.2 Matrix Metalloproteases

Members of the matrix metalloprotease gene family (Birkedal-Hansen et al. 1993) are both AP1 and estrogen responsive (Rajabi et al. 1991; Rodgers et al. 1994). They are involved in tissue remodeling in breast and uterus. ER action at the native collagenase gene promoter requires the AP1 site (Fig. 4). Inspection of other metalloprotease promoters showed that they also contained AP1 sites, but no apparent consensus palindromic EREs. Thus, metalloproteases are candidate genes that could be regulated by the ER/AP1 pathway.

7.9.3 Composite Elements in the Metalloprotease Gene Family? The Gelatinase Proximal Promoter

In addition to a consensus AP1 site, the gelatinase proximal promoter contains several copies of a variant AP1 site which was first recognized in the ovalbumin promoter (ov-PE; Tora et al. 1988). This element resembles an ERE half site, but also binds AP1 proteins, and was the

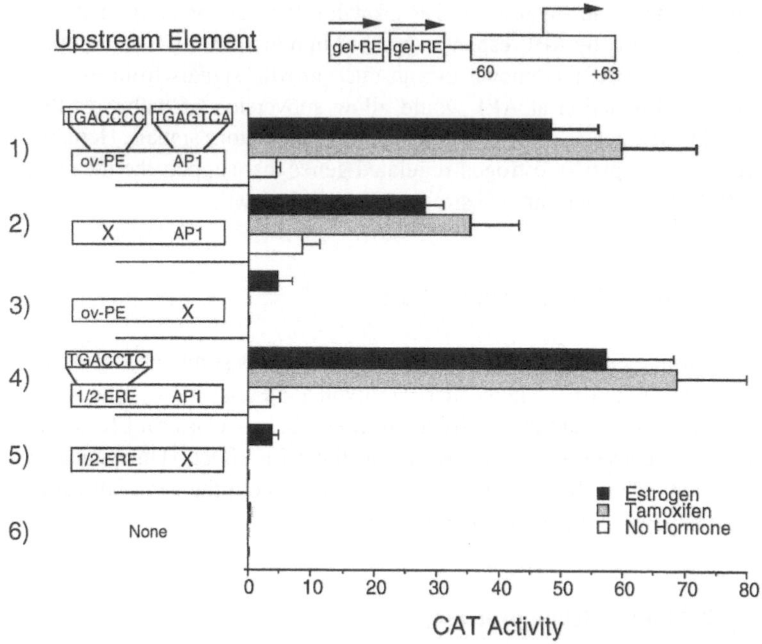

Fig. 7. Sequence of the gelatinase variant AP1 site (*Gel-PE*). Ovalbumin and consensus AP1 sites are marked. Behavior of the gel-PE is shown, along with various mutations, in Chinese hamster ovary cells with stably transfected estrogen receptor. The reporter genes contain native gel-PE (*1*), gel-PE consensus AP1 site alone (*2*), ovalbumin promoter (*ov-PE*) site alone (*3*), a gel-PE with ov-PE site mutated to resemble a consensus estrogen response element (*ERE*) (*4*), ERE half site alone (*5*), and core collagenase promoter (*6*). *Gel-RE*, gelatinase response element

first example of an alternate ERE (Gaub et al. 1990). One ov-PE site lay immediately upstream of the gelatinase consensus AP1 site.

We tested the properties of this element (gel-PE, consisting of both the ov-PE site/consensus AP1 site) upstream of the collagenase core promoter. We found that a dimerized gel-PE elicited large tamoxifen and estrogen responses (Fig. 7). Removing the ov-PE site reduced ER

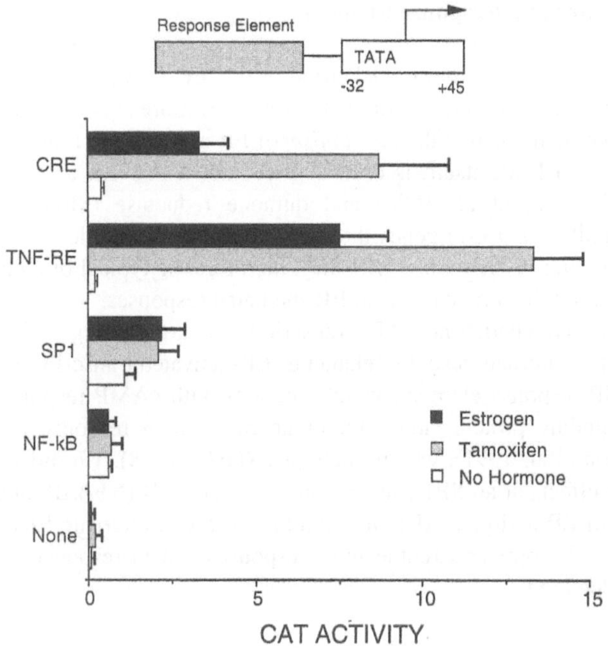

Fig. 8. Estrogen receptor (ER) action at alternate response elements (RE). Reporter genes were transfected into HeLa cells along with ER expression vector. We asked whether CREs, tumor necrosis factor-REs (*TNF-RE*), SP1 or NF-kB sites, placed upstream of the herpes simplex virus-TK promoter (*HSV-TK*) tata box, would allow ER action. *CRE*, cyclic AMP response element; *NF-κB*, nuclear factor-κB

effects. Mutation of the AP1 site, however, leaving only the ov-PE site, abolished both basal activity and tamoxifen effects. Thus, in the context of the gelatinase element, the ov-PE site magnifies ER effects at the consensus AP1 site.

We also examined the behavior of a similar element in which we had altered the sequences downstream of the ov-PE site so that it resembled an ERE half site. Estrogen and tamoxifen responses were not altered by this mutation. This raises the possibility that composite ERE/AP1 elements could be targets for ER action.

7.9.4 Alternate Response Elements

In its strongest form, our model proposes that ER activates AP1-responsive genes which, in turn, stimulate growth. A more subtle variation of the same model is that the *mechanism* of ER action at AP1 sites, rather than the AP1 site itself, is crucial. It is known that ER activates the TGFβ3 (Yang et al. 1996) and quinone reductase (Montano and Katzenellbogen 1997) genes through a mechanism that does not need ERE binding or AP1 sites. AP1 may, therefore, be typical of a class of transcription factors that permit ER-mediated responses.

We asked whether non-AP1 transcription-factor binding sites would also act as alternate response elements. ER-activated transcription from a cAMP response element, which interacts with cAMP response element binding protein, and a tumor necrosis factor response element (Leitman et al. 1991), which binds jun/ATF2 (Fig. 8). We did not see similar effects at an SP1 site or a nuclear factor-κB (NF-κB) site. The nature of ER action at AP1 sites and the variety of alternate EREs will doubtless become apparent as more response element-responsive genes are understood.

7.10 Summary and Discussion

We began this study looking for simple reporters that might help us model tamoxifen effects. We found that ER action at AP1-responsive reporter genes correlated well with tamoxifen and estrogen effects on growth. Equally importantly, ER effects on standard ERE-responsive reporter genes poorly predicted ER effects on growth. Indeed, in breast cells, AF1 activity on simple ERE-responsive reporters showed a negative correlation with tamoxifen effects. We therefore suggested that studying ER action at AP1 might be a good way of predicting antiestrogen behavior.

To test whether the AP1 paradigm would have predictive value, we analyzed the behavior of other well characterized antiestrogens with AP1-responsive reporters. We found that nonclassical antiestrogen effects of GR and PR were reiterated at AP1 sites. We also found that raloxifene, which has little uterotropic activity, had little activity at AP1 sites. Because the present generation of pure antiestrogens also activate

AP1, we suspect that they will turn out to be agonists in conditions in which they do not down regulate ER protein. We also predict that ERβ, which strongly enhances AP1 activity in response to antiestrogens, will prove to be an important player in antiestrogen action in vivo.

During the past few years it has emerged that several genes, some of which are clearly involved in growth response, are regulated by ER through AP1 sites, or other alternate response elements. This raises an important question: To what extent do alternate elements, such as AP1, contribute to observed physiological effects of estrogen and antiestrogens? To address this question we are looking for ER mutations that specifically block estrogen or antiestrogen action in either the classical or AP1 pathways. We are also trying to understand how ER acts at AP1 sites in tissue culture. Our results suggest, at the very least, that understanding ER action at alternate response elements will help us to understand how antiestrogens work. If ER action at AP1 sites does turn out to be an important mechanism of ER growth regulation, then a fuller understanding of ER/AP1 interactions will allow us to design the next generation of antiestrogens and hormone replacement therapies.

References

Ali S, Metzger D, Bornert JM, Chambon, P. (1993) Modulation of transcriptional activation by ligand-dependent phosphorylation of the human oestrogen receptor A/B region. EMBO J 12:1153–60

Anonymous (1992) Breast cancer prevention trial will recruit 16 000 women. Oncology (Williston Park) 6:57–58

Anzick SL, Kononen J, Walker RL, Azorsa DO, Tanner MM, Guan XY, Sauter G, Kallioniemi OP, Trent JM, Meltzer PS (1997) AIB1, a steroid receptor coactivator amplified in breast and ovarian cancer. Science 277:965–968

Berry M, Metzger D, Chambon P (1990) Role of the two activating domains of the oestrogen receptor in the cell-type and promoter-context dependent agonistic activity of the anti-oestrogen 4-hydroxytamoxifen. EMBO J 9:2811–2818

Birkedal-Hansen H, Moore W, Bodden M, Windsor L, Birkedal-Hansen B, DeCarlo A, Engler J (1993) Matrix metalloproteases; a review. Oral Biol Med 4:197–250

Cavaillès V, Dauvois S, Danielian PS, Parker MG (1994) Interaction of proteins with transcriptionally active estrogen receptors. Proc Natl Acad Sci USA 91:10009–10013

Chen H, Lin RJ, Schiltz RL, Chakravarti D, Nash A, Nagy L, Privalsky ML, Nakatani Y, Evans RM.(1997) Nuclear receptor coactivator ACTR is a novel histone acetyltransferase and forms a multimeric activation complex with P/CAF and CBP/p300. Cell 90:569–580

Dauvois S, Danielian PS, White R, Parker, MG (1992) Antiestrogen ICI 164,384 reduces cellular estrogen receptor content by increasing its turnover. Proc Natl Acad Sci USA 89:4037–41

Dotzlaw H, Leygue E, Watson PH, Murphy LC (1997) Expression of estrogen receptor-beta in human breast tumors. J Clin Endocrinol Metab 82:2371–2374

Gaub MP, Bellard M, Scheuer I, Chambon P, Sassone-Corsi P (1990) Activation of the ovalbumin gene by the estrogen receptor involves the fos-jun complex. Cell 6:1267–1276

Green S (1990) Modulation of oestrogen receptor activity by oestrogens and anti-oestrogens. J Steroid Biochem Mol Biol 37:747–751

Gronemeyer H, Benhamou B, Berry M, Bocquel MT, Gofflo D, Garcia T, Lerouge T, Metzger D, Meyer ME, Tora L, Vergezac A and Chambon P (1992) Mechanisms of Antihormone Action. J Steroid Biochem Mol Biol 41:217–221

Hong H, Kohli K, Trivedi A, Johnson DL, Stallcup M R (1996) GRIP1, a novel mouse protein that serves as a transcriptional coactivator in yeast for the hormone binding domains of steroid receptors. Proc Natl Acad Sci USA 93:4948–4952

Horwitz KB (1994) How do breast cancers become hormone resistant? J Steroid Biochem Mol Biol 49:295–302

Huynh HT, Pollak M (1993) Insulin-like growth factor I gene expression in the uterus is stimulated by tamoxifen and inhibited by the pure antiestrogen ICI 182780. Cancer Res 23:5585–5588

Jamil A, Croxtall JD, White J O (1991) The effect of anti-oestrogens on cell growth and progesterone receptor concentration in human endometrial cancer cells (Ishikawa). J Mol Endocrinol 6:215–21

Jordan VC (1992) The strategic use of antiestrogens to control the development and growth of breast cancer. Cancer 70:977–982

Katzenellenbogen BS, Korach KS (1997) Editorial: a new actor in the estrogen receptor drama – enter ER beta. Endocrinology 138:861–862

Katzenellenbogen JA, O'Malley BW, Katzenellenbogen BS (1996) Tripartite steroid hormone receptor pharmacology: interaction with multiple effector sites as a basis for the cell- and promoter-specific action of these hormones. Mol Endocrinol 10:119–131

Kedar RP, Bourne TH, Powles TJ, Collins WP, Ashley SE, Cosgrove DO, Campbell S (1994) Effects of tamoxifen on uterus and ovaries of postmenopausal women in a randomised breast cancer prevention trial. Lancet 343:1318–21

Kuiper GGJM, Enmark E, Pelto-Huikko M, Nilsson S, Gustafsson JA (1996) Cloning of a novel estrogen receptor expressed in rat prostate and ovary. Proc Natl Acad Sci USA 93:5925–5930.

Leitman DC, Ribeiro RCJ, Mackow ER, Baxter JD, West BL (1991) Identification of a tumor necrosis factor-responsive element in the tumor necrosis factor alpha gene. J Biol Chem 266:9343–9346

Li H, Gomes PJ, Chen JD (1997) RAC3, a steroid/nuclear receptor-associated coactivator that is related to SRC-1 and TIF2. Proc Natl Acad Sci USA 94:8479–8484

McDonnell DP, Clemm DL, Hermann T, Goldman ME, Pike JW (1995) Analysis of estrogen receptor function in vitro reveals three distinct classes of antiestrogens. Mol Endocrinol 9:659–69

Montano MM, Katzenellenbogen BS (1997) Identification of a novel cis-acting element in the promoter of an estrogen-responsive gene that modulates sensitivity to hormone and antihormone. Proc Natl Acad Sci USA 94:2581–2585

Morrow M, Jordan VC (1993) Molecular mechanisms of resistance to tamoxifen therapy in breast cancer. Arch Surg 128:1187–1191

Mosselman S, Polman J, Dijkema R (1996) ER beta: identification and characterization of a novel human estrogen receptor. FEBS Lett 392:49–53

Onate SA, Tsai SY, Tsai MJ, O'Malley BW (1995) Sequence and characterization of a coactivator for the steroid hormone receptor superfamily. Science 270:1354–7

Onoe Y, Miyaura C, Ohta H, Nozawa S, Suda T (1997) Expression of estrogen receptor beta in rat bone. Endocrinology 138:4509–4512

Paech K, Webb P, Kuiper GGJM, Nilsson S, Gustafsson J, Kushner PJ, Scanlan TS (1997) Differential ligand activation of estrogen receptors ER alpha and ER beta at AP1 sites. Science 277:1508–1510

Parker MG (1990) Mechanisms of action of steroid receptors in the regulation of gene transcription. J Reprod Fertil 2:717–720

Philips A, Chalbos D, Rochefort H (1993) Estradiol increases and anti-estrogens antagonize the growth factor-induced activator protein-1 activity in MCF7 breast cancer cells without affecting c-fos and c-jun synthesis. J Biol Chem 268:14103–14108

Rajabi M, Solomon S, Poole AR (1991) Hormonal regulation of interstitial collagenase in the uterine cervix of pregnant guinea pig. Endocrinology 128:863–871

Rodgers WH, Matrisian LM, Guidice LC, Dsupin B, Cannon P, Svitek C, Gorstein F, Osteen KG (1994) Patterns of metalloprotease expression in cycling

emdometrium imply differential functions and regulation by steroid hormones. J Clin Invest 94:946–953
Tora L, Gaub MP, Mader S, Dierich A, Bellard M, Chambon P (1988) Cell-specific activity of a GGTCA half-palindromic oestrogen-responsive element in the chicken ovalbumin gene promoter. EMBO J 12:3771–3778
Torchia J, Rose DW, Inostroza J, Kamei Y, Westin S, Glass CK, Rosenfeld MG (1997) The transcriptional co-activator p/CIP binds CBP and mediates nuclear-receptor function. Nature 387:677–684
Tremblay GB, Tremblay A, Copeland NG, Gilbert DJ, Jenkins NA, Labrie F, Giguere V (1997) Cloning, chromosomal localization, and functional analysis of the murine estrogen receptor beta. Mol Endocrinol 11:353–365
Uht RM, Anderson CM, Webb P, Kushner PJ (1997) Transcriptional activities of estrogen and glucocorticoid receptors are functionally integrated at the AP-1 response element. Endocrinology 138:2900–2908
Umayahara Y, Kawamori R, Watada H, Imano E, Iwama N, Morishima T, Yamasaki Y, Kajimoto Y, Kamada T (1994) Estrogen regulation of the insulin-like growth factor I gene transcription involves an AP-1 enhancer. J Biol Chem 269:16433–16442
Voegel JJ, Heine MJS, Zechel C, Chambon P, Gronemeyer H (1996) TIF2, a 160 kDa transcriptional mediator for the ligand-dependent activation function AF-2 of nuclear receptors. EMBO J 15:3667–3675
Wakeling AE (1993) The future of new pure antiestrogens in clinical breast cancer. Breast Cancer Res Treat 25:1–9
Webb P, Lopez GN, Uht RM, Kushner PJ (1995) Tamoxifen activation of the estrogen receptor/AP-1 pathway: potential origin for the cell-specific estrogen-like effects of antiestrogens. Mol Endocrinol 9:443–456
Yang NN, Venugopalan M, Hardikar S, Glasebrook A (1996) Identification of an estrogen response element activated by metabolites of 17-beta-estradiol and raloxifene. Science 273:1222–1225
Zhou F, Bouillard B, Pharaboz-Joly MO, Andre J (1989) non-classical antiestrogenic actions of dexamethasone in variant MCF7 breast cancer cells in culture. Mol Cell Endocrinol 66:189–197

8 A Structural View of the Retinoid Nuclear Receptors

B.P. Klaholz and D. Moras

8.1 The Nuclear Receptor Superfamily

Cellular signaling uses two main pathways, the membrane-associated and the nuclear receptors (NRs). The characteristic of the first is the use of water-soluble ligands which do not cross the membrane. Ligand binding at the cell surface induces, e.g., autophosphorylation of epidermal growth factor (EGF) receptor, starting a cascade of serine- or tyrosine-kinases, finally transducing the signal to the nucleus. NRs work very differently. Located in the nucleus they control the activity of their target genes directly by binding to specific DNA sequences called hormone response elements. They are activated by hydrophobic ligands, like the steroid hormones and retinoids, which reach their receptor in the cytoplasm or the nucleus by crossing the lipid bilayer of the cell mem-

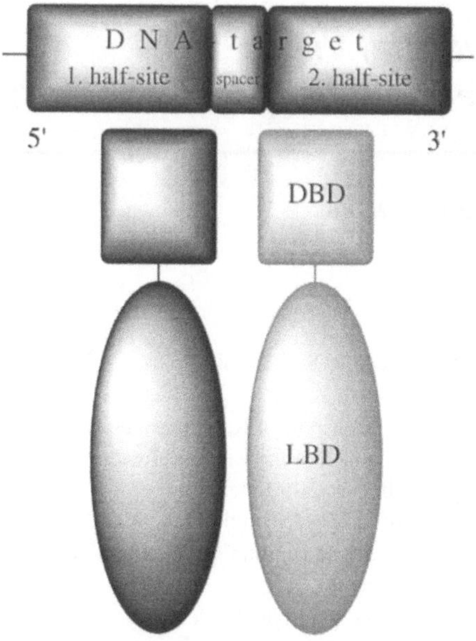

Fig. 1. Characteristic heterodimerization of the nuclear receptors for retinoic acid (*RAR*), thyroid hormone (*TR*), and vitamin D (*VDR*) (in *grey*) with their common partner, retinoid X receptor (*RXR*), (in *black*), mediated by the appropriate DNA target. Note that in general RXR occupies the 5'-position. *DBD*, DNA-binding domain

brane. Some of these ligands, such as retinoic acid, need to be metabolically modified, others are completely synthesized in the cell, as prostaglandins.

The superfamily of nuclear receptors includes those for steroid hormones [estrogen receptor (ER), glucocorticoid receptor (GR), among others], vitamin D, thyroid hormone, and the retinoids [the receptors are VDR, TR, retinoic acid receptor (RAR), and retinoid X receptor (RXR), respectively], which are vitamin A metabolites. All these different ligands bind selectively to their target receptor, and thereby regulate transcription. Combinations of different ligands in homo- or heterodimeric receptor complexes illustrate their important role in development and cell differentiation (reviewed in Mangelsdorf et al. 1995; Mangelsdorf and Evans 1995). Additionally, the so-called orphan receptors are either constitutively active or their ligands are not yet known. The investigation of orphan receptors is of particular interest, as ligand binding appears to have been acquired during evolution (Escriva et al. 1997).

NRs exhibit a modular structure with regions A through F, corresponding to distinct structural domains. The A-B regions contain the ligand-independent activation function 1 (AF-1). Region C is the DNA-binding domain (DBD), comprising two highly conserved zinc binding motifs and the DBD dimerization surface. Region E is the ligand-binding domain (LBD), which harbors the main regulation functions, namely the ligand-binding site, the ligand-dependent activation function 2 (AF-2), and the LBD dimerization surface.

NRs can form different kinds of dimers which bind to DNA, e.g., the ER forms homodimers, whereas RAR, TR, and VDR mainly form heterodimers with RXR. This is schematically shown in Fig. 1.

8.2 Retinoid Receptors

The DBD of the retinoid receptors contains a 66 residue core and the LBD comprises approximately 225 residues. These domains are linked by a short hinge region, which, by analogy to the steroid hormone receptors, may work as a nuclear translocation signal (Ylikomi et al. 1992; Giguere 1994).

The three subtypes α, β, and γ of RAR and RXR are encoded by different genes, and their multiple isoforms, which differ in their N-terminal A region, are generated by differential promoter usage and alternative splicing. RARs are activated by both 9-*cis* and all-*trans* retinoic acid (9C-RA and AT-RA, respectively), whereas the RXRs are selective for 9C-RA (Levin et al. 1992; Allenby et al. 1993).

Although RARs and RXRs can form homodimers, the formation of heterodimers increases the affinity for the cognate response elements leading to the formation of anisotropic receptor–DNA complexes (Zechel et al. 1994b; Kurokawa et al. 1994) (see Fig. 1). This polarity is due to the formation of dimers on direct repeats, whereas complexes on palindromes or everted repeats exhibit C_2-symmetry (Gronemeyer and Moras 1995). The core of the RAR and RXR response elements (RAREs and RXREs) is a direct repeat (DR) of the hexanucleotide PuGGTCA, spaced by 1, 2, or 5 nucleotides. With the exception of RAR/RXR dimers on DR1, RXR occupies the 5'-position in heterodimers with RAR (DR2, DR5), VDR (DR3), and TR (DR4), or certain orphan receptors (Lee et al. 1993; Mader et al. 1993; Zechel et al. 1994a; Rastinejad et al. 1995; Mangelsdorf et al. 1995).

8.3 Metabolism and Transport of Retinoids and Their Precursors

Vitamin A (retinol) is the common precursor of retinal and retinoids. Retinol is transported in the serum by the retinol-binding protein (RBP), and its cell uptake appears to be mediated by a putative membrane RBP receptor. Retinol may either be stored as an acetylester or oxidized to retinal (see Fig. 2). In the first case, a complex of retinol and the cellular RBP (CRBP-II) serves as a substrate for esterification by the lecithin:retinol acetyltransferase. In the second case, a CRBP-I complex is utilized by retinol-dehydrogenase isozymes for the conversion to retinal, followed by the oxidation to the corresponding acid by a cytosolic retinal dehydrogenase. The resulting RA is still mainly hydrophobic and can be stored or transported by cellular RA-binding proteins (CRABP), or finally bound by retinoid receptors before being catabolically inactivated by cytochrome P450 to increasingly polar compounds (reviewed in Giguere 1994; Napoli 1996).

It is interesting to note that on the one hand the promoter of the CRABP-II gene contains both a RARE and a RXRE, and on the other hand the promoters of the RAR genes also have a RARE, leading to an autoregulation of the retinoid signal. This complex interplay shows that retinoid receptors are part of a metabolite-controlled signaling system.

Fig. 2. The metabolism of retinol (*ROH*), leading to the retinoids, the ligands for the retinoid receptors *RAR* and RXR (adapted from Giguere 1994). *RBP*, retinol-binding protein; *CRBP*, cellular RBP; *L-RAT*, lecithin:retinol acetyltransferase; *REH*, retinyl ester hydrogenase; *RAL*, retinal; *CRABP*, cellular retinoic acid binding protein; *AT-RA*, all-*trans* retinoic acid ▶

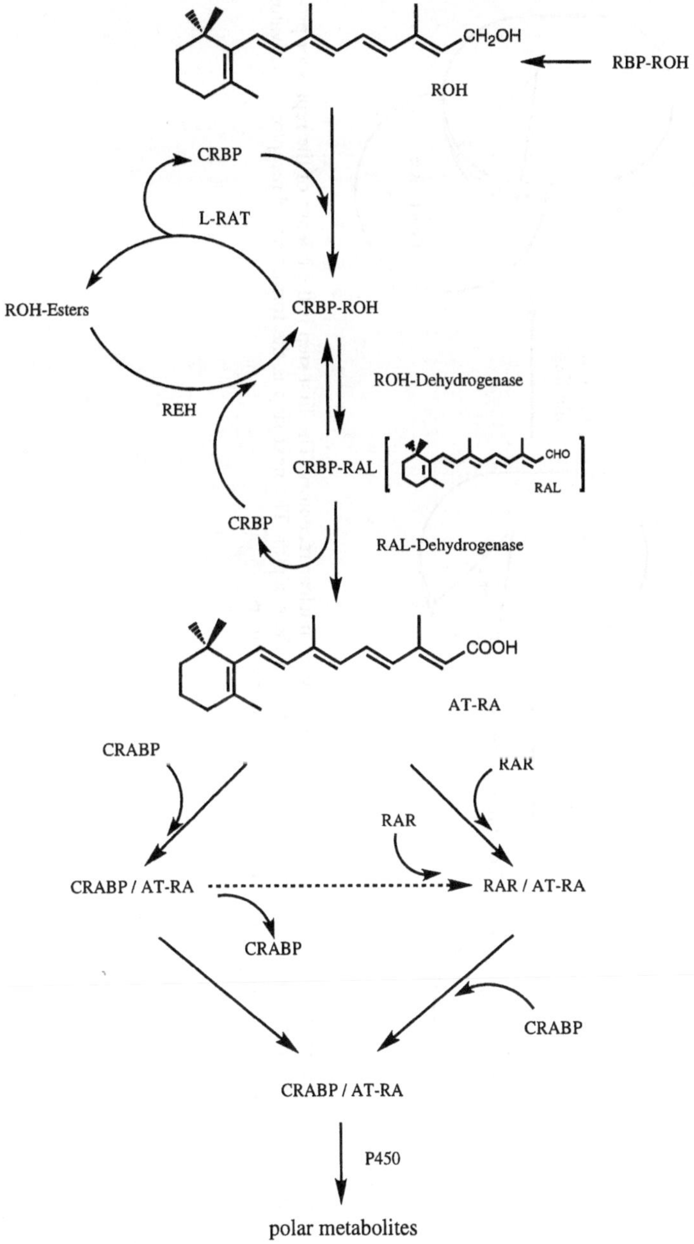

Fig. 2. Legend see p. 144

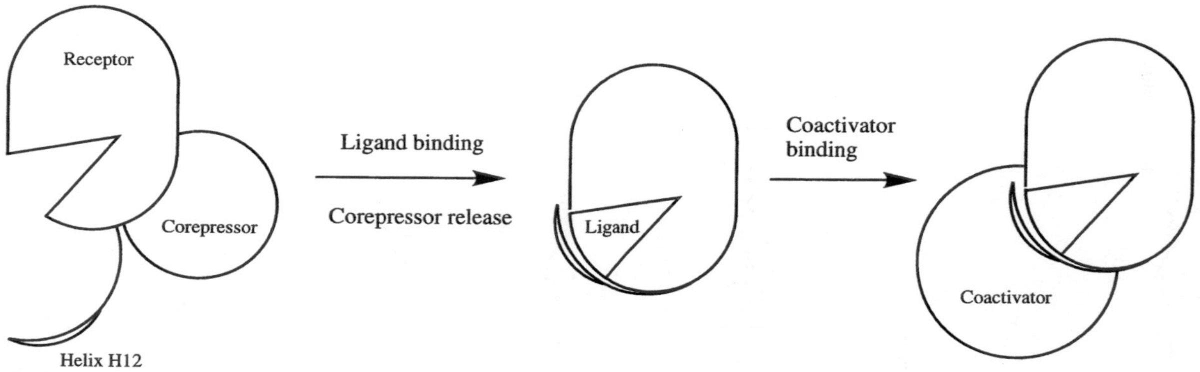

Fig. 3. Two-step mechanism of ligand-induced activation of nuclear receptors. The first step is the release of the repressor protein upon ligand binding, leading to the activated state of the receptor. The next step is the formation of receptor–coactivator complexes which in turn interact with the transcription machinery

8.4 Activation Mechanism of the Retinoid Receptors

The switch from inactive to active receptors is correlated to different molecular events. Non-liganded RARs and RXRs are inactive because they form complexes with repressor proteins like the silencing mediator for RARs and TRs (SMRT) (Chen and Evans 1995; Chen et al. 1996) and the nuclear receptor corepressor (N-CoR) (Hörlein et al. 1995). Ligand binding induces a conformational change of the receptor leading to the release of the corepressor (Fig. 3; reviewed in Horwitz et al. 1996). This now adopted conformation represents the activated state of the receptor, but different intermediary proteins like the steroid receptor coactivator (SRC-1) (Oñate et al. 1995; McInerney et al. 1996) and the transcriptional intermediary factor (TIF2) (Voegel et al. 1996) can bind to the activated receptor and create a link to the basal transcription machinery. This type of activation works essentially through the activation function AF-2 of the LBD and controls transcription.

Another ligand-independent activation mode is the phosphorylation of AF-1 of the B-domain. It has been recently shown that Ser77 of RAR is phosphorylated by cdk7 (Rochette-Egly et al. 1997), the kinase of the CAK complex (comprising cdk7, cyclin H, MAT-1), which is part of the general transcription and repair factor TFIIH.

8.5 Retinoids and the Need for Selectivity

Several epidemiological studies have shown that people with lower dietary vitamin A intake are at higher risk of developing cancer (Lotan 1997 and references therein). The use of retinoids in the treatment of skin diseases including psoriasis, acne, and skin cancer, as well as the treatment of chronic myelogenous leukemia, cervical cancer, and in cancer chemoprevention (Tallman and Wiernik 1992; Giguere 1994; Niles 1995) is related to the multiple effects of RA on morphogenesis, differentiation, and homeostasis, regulating key steps in cell growth, development, and embryogenesis (De Luca 1991; Kastner et al. 1995; Chambon 1996). Cancer development is associated with dysregulation of cell proliferation, aberrant differentiation, and the loss of ability or decreased tendency to undergo apoptosis.

RAR and RXR play a key role in mediating the retinoid effects on gene expression and the above-mentioned processes. The RAR subtypes are activated by both retinoic acid isomers, whereas the RXRs are selective for 9C-RA. Therefore, specific compounds should act selectively in gene regulation and be at the same time less prone to side effects, e.g., the teratogenicity of retinoic acid. The discovery of RAR subtype-selective compounds further emphasizes the need for structural information. Recently, the crystal structures of the LBDs of RXR (Bourguet et al. 1995) and RAR (Renaud et al. 1995; Klaholz et al. 1998) have been solved in our laboratory. These structures have enabled a general sequence alignment of all known NR LBDs (Wurtz et al. 1996) and have led to both an insight into the ligand-binding mechanism and the ligand specificity of the retinoid receptors. Drug design based on these crystal structures is now possible and may lead to more selective compounds.

8.6 The Structure of the DNA-Binding Domain

The DBD structures of the GR, ER, and RXR/TR have been described (Luisi et al. 1991; Schwabe et al. 1993; Rastinejad et al. 1995). The highly conserved motif in these domains is a helix-loop-helix stabilized by two zinc ions, each being coordinated tetrahedrally by four cysteines. The functional DNA recognition unit for most NRs is a dimer. Helix 1 of each unit, the recognition helix, interacts with the major groove of the DNA. In the complex of the RXR/TR DBD dimer bound to the DR4 response element (Rastinejad et al. 1995), TR exhibits an additional C-terminal helix extending across the minor groove of the DNA, which by analogy with VDR and GR contains the sequence for the nuclear localization signal. The RXR DBD occupies the upstream half-site AGGTCA, in agreement with the above-mentioned biochemical data (Lee et al. 1993; Mader et al. 1993; Zechel et al. 1994a). The tandem arrangement of the half-sites in the DR4 imposes a head-to-tail orientation of the DBDs on the DNA. This leads to an anisotropic complex, which is formed due to protein–protein interaction, favored exclusively in this orientation.

Note that in the case of the steroid receptors the palindromic DNA repeats lead to a symmetric head-to-head arrangement, seen in the GR and ER DBD–DNA complexes (Luisi et al. 1991; Schwabe et al. 1993).

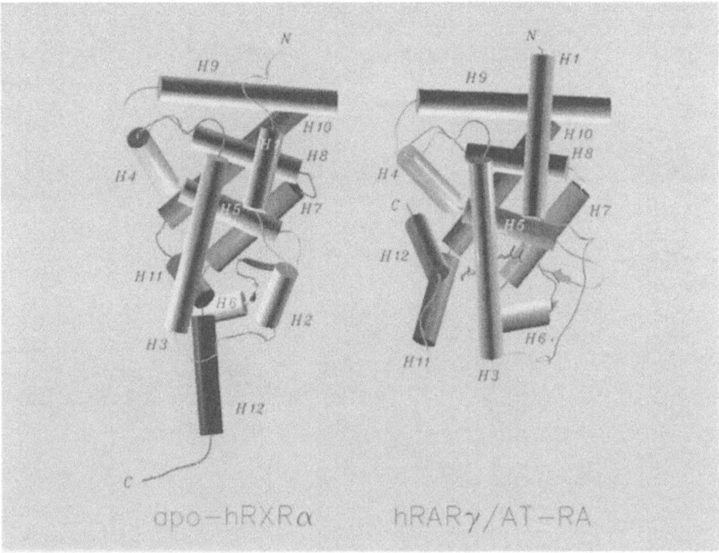

Fig. 4. The ligand-binding domain of the retinoid receptors: comparison of the unliganded human retinoid receptor α (*apo-hRXRα*) (Bourguet et al. 1995) with the human retinoic acid receptor γ (*holo-hRARγ*) bound to all-*trans* retinoic acid (*AT-RA*) (Renaud et al. 1995), illustrating the common fold, an antiparallel α-helical sandwich (adapted from Renaud et al. 1995). The numbering of the helices (*H1–H12*) and the N- and C-termini (*N/C*) are indicated. All pictures were generated using the program Setor (Evans 1993)

The common feature of NR DBDs not to dimerize on themselves indicates that the interaction between the DBDs is mediated by the DNA target (Luisi et al. 1991). The spacer length between the half-sites is responsible for the selectivity by favoring only the association of the correct DBD pairs.

8.7 LBD Structure, Ligand Binding, and Activation of Retinoid Receptors

The crystal structures of the LBDs of apo-hRXRα (Bourguet et al. 1995) and hRARγ bound to AT-RA (Renaud et al. 1995) and 9C-RA

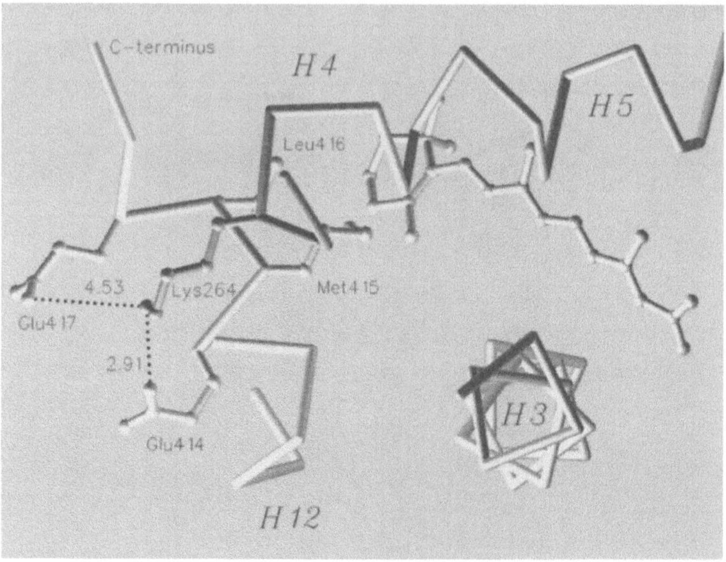

Fig. 5a,b. Detailed view of the ligand-binding domain of the human retinoid receptor γ bound to all-*trans*-retinoic acid. **a** The EMLE-motif of the amphipathic transactivation helix *H12*, where *Glu414* provides a strong salt bridge to *Lys264* of helix *H4*, stabilizing the conformation of *H12*, and the methionine and leucine residues oriented towards the ligand. **b** AT-RA bound in the ligand-binding pocket, exhibiting hydrophobic contacts with the β-ionone moiety and a hydrophilic network of salt bridges and hydrogen bonds to the carboxylate group (adapted from Renaud et al. 1995). H12 is omitted for clarity and would close the entry being salt-bridged to the indicated *Lys264*

(Klaholz et al. 1998) have been determined at atomic resolution (2.7, 2.0, and 2.4Å, respectively). The first result of these studies was to show that both LBDs of RXR and RAR exhibit a novel common fold, an antiparallel α-helical sandwich (Fig. 4). This common fold has been used for a general sequence alignment of the NR LBDs (Wurtz et al. 1996), which has been confirmed by the LBD structures of TR and ER (Wagner et al. 1995; Brzozowski et al. 1997). In Fig. 4 the missing N-terminal flanks of RAR and RXR are connected to the DBD and in RAR the C-terminus extends to the F-domain. The first helix at the N-terminus of the RAR LBD is H1. The residues of RAR which corre-

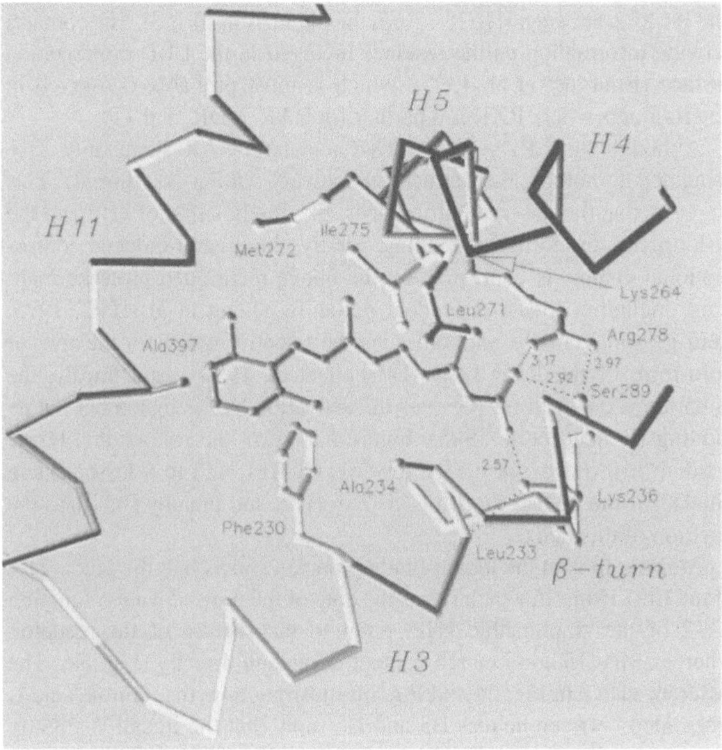

Fig. 5h. Legend see. p. 150

spond to H2 of RXR form a loop between H1 and H3. H3 is the longest helix and is bent against the receptor core in the holo-RAR. Helices H5, H8, and H9 form the middle layer of the sandwich. A short β-turn is located between H5 and H6. Helix H3, the β-turn, H4, H5, H6, H11, and H12 participate in the formation of the ligand-binding pocket. In the apo-RXR the C-terminal H12 and the Ω-loop between H2 and H3 point into the solvent, leading to a relatively loose structure sensitive to proteolysis, whereas the more compact holo-RAR LBD is clearly more stable (Leid 1994; Keidel et al. 1994). In the crystal structure, the apo-RXR is seen as a homodimer, where H10 – and to a lesser extent H9 and the loop between H7/H8 – form the interaction surface. The contacts provide information on the residues involved in the LBD dimerization surface (Bourguet et al. 1995), which is most probably conserved in heterodimers where RXR is a partner for RAR, VDR, and TR.

Although the LBDs of apo-RXR and holo-RAR share only 27% sequence homology, their structures are very similar (see Fig. 4). This suggests that the observed differences, like the position of H12 and the Ω-loop, are due to ligand binding. Firstly, this ligand-induced conformational change is confirmed by the above-mentioned protease mapping, including ligand-dependent gel-shifts (Leng et al. 1993, 1995; Leid 1994). Secondly, antibodies can be selective either for the apo- or holo-form of the RAR LBD (Driscoll et al. 1996). And thirdly, the mutation K264 A in H4 prevents the activation by the ligand but not its binding (Renaud et al. 1995). Figure 5a shows the role of this lysine residue: it forms a salt bridge towards Glu414, and to a lesser extent Glu417 of the EMLE-motif of AF-2 in H12, and thereby stabilizes the position of this helix.

A closer look at the ligand-binding pocket shows that the glutamates of the EMLE-motif, which forms the core of the transactivation function AF-2 of the amphipathic H12, point to the surface of the receptor, whereas Met415 and Leu416 contact the ligand directly (Fig. 5a). The different views in Fig. 5a and Fig. 5b illustrate how the retinoic acid is embedded between helices H3 and H5; note that the ligand is slightly bent and is twisted by 43° between the β-ionone moiety and the tetraene chain, which is due to the van der Waals contacts with the surrounding residues. The β-ionone moiety interacts with hydrophobic isoleucines, leucines, phenyalanines, alanines, and methionines. The carboxylate group, however, is involved in a hydrophilic network, including salt bridges to Arg278 and Lys236, and hydrogen bonds to Ser289 (β-turn) and the main chain carbonyl group of Leu233 (H3) mediated by a water molecule, or Nζ of Lys236 in an alternative conformation replacing the water molecule (Renaud et al. 1995).

Renaud et al. (1995) have proposed a mechanism of ligand entry based on an electrostatic field guidance. The negatively charged ligand appears to be attracted into the ligand-binding pocket by a positively charged cluster including Arg278, Lys236, Lys229, Lys240, and Arg274. Figure 6 shows the result of an electrostatic field calculation with the programme GRASP (Nicholls et al. 1991) where the ligand was

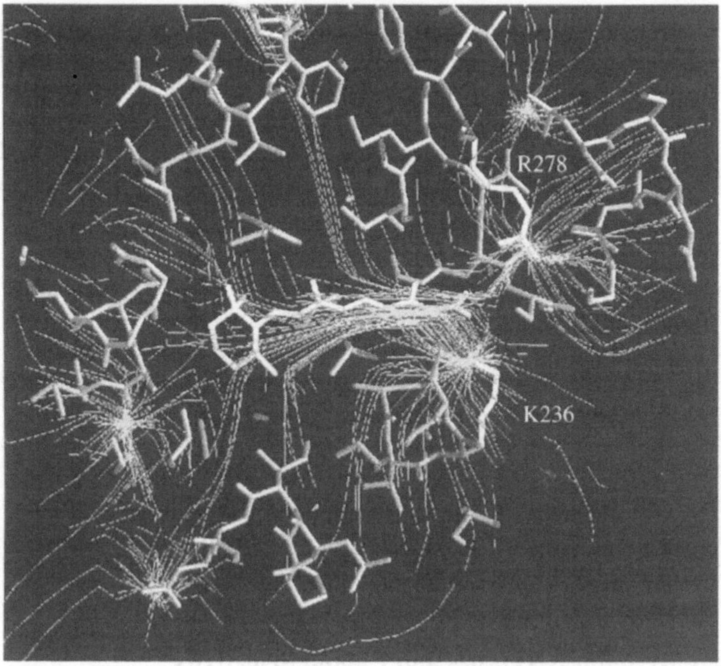

Fig. 6. View of the ligand-binding pocket with the electrostatic field lines calculated with the program GRASP, ending on the surface of the cavity and oriented from negative to positive charges. The ligand was omitted for the calculation and added for the figure (adapted from Renaud et al. 1995)

omitted: the bundle of field lines leads to the charged cluster through the cavity, indicating the orientation of the ligand (added in the figure after the calculation). As Lys264 is located at the entrance of the cavity it could help to guide the ligand during its entry, before forming the salt bridge with Glu414. At the same time, the receptor undergoes a conformational change, mainly involving the refolding of the Ω-loop and H12: compared to the apo-RXR, the Ω-loop undergoes a hinged-lid motion underneath the receptor (seen in the holo-RAR, Fig. 4), and is subsequently fixed by a salt bridge between Arg219 and Asp309 (H7). The C-terminal helix H12 nestles against the receptor and closes the entry of the receptor after ligand binding, being stabilized by the salt bridge between Glu414 and Lys264 of H4 (see Fig. 5a). Simulations using molecular dynamics seem to confirm the proposed mechanism (A. Blondel et al., in preparation). For these calculations the structures of the apo-hRXRα and the AT-RA/hRARγ complex were used as models of the apo- and holo-form, and homology modeling of the RXR provided an apo-RAR model. The calculations were run for a ligand exit and are consistent with the proposed mechanism for a ligand entry and the major conformational change mentioned earlier. The mechanism may be different for uncharged ligands of other NRs.

The conformation of H12 in the holo-RAR is characteristic of the active form of the receptor, firstly because only the holo-receptor is transcriptionally active, secondly the mutation K264 A almost abolishes transactivation without impairing ligand and DNA binding, and thirdly coactivators appear to interact only with the liganded receptor. This indicates that the conformational change generates a new surface enabling the formation of receptor–coactivator complexes crucial for transcription regulation, and also explains the ligand-induced release of corepressors (see Fig. 3).

8.8 Retinoid Receptors and Ligand Selectivity

Three types of selectivity have to be considered for NR ligands. Firstly, they have preferences for a family, e.g., steroid hormone versus retinoid receptors. Secondly, there is selectivity for subfamilies, e.g., RAR versus RXR, and last but not least for their subtypes, i.e., RARα, β, and γ.

The subtype selectivity of synthetic retinoids is mainly due to residues in the ligand-binding pocket. The sequence alignment of RARα, β and γ shows that only three residues differ in the pocket (Renaud et al. 1995; Wurtz et al. 1996) and their mutation induces specificity switches (M. Gehin et al., in preparation). In RARα the presence of Ser232, Ile270, and Val395 probably leads to a smaller cavity compared to RARβ (Ala225, Ile263, Val388) and RARγ (Ala234, Met272, Ala397) and may in part explain the RARα selectivity of the relatively small AM580 (Bernard et al. 1992). Figure 7 shows a few known synthetic retinoids, exhibiting RAR or RXR specificity and additionally RAR-subtype selectivity (Bernard et al. 1992; Dawson et al. 1995; Hembree et al. 1996). The ligands contain on the one hand a hydrophobic moiety,

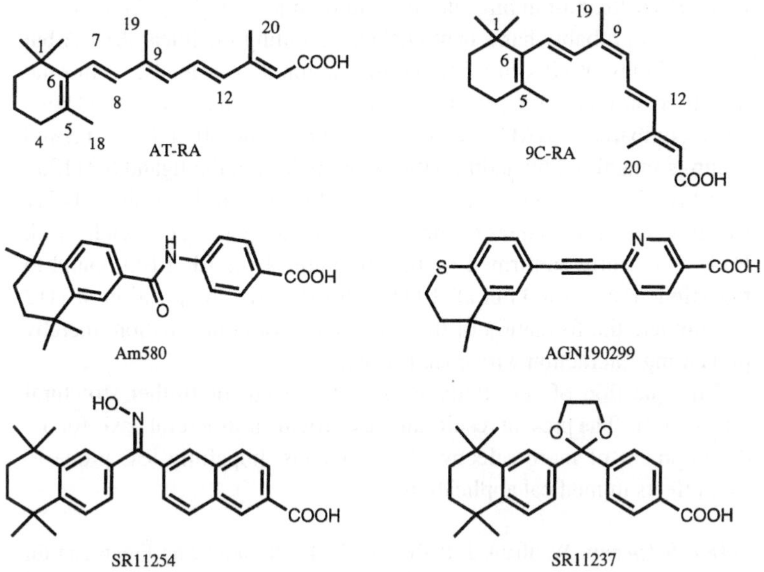

Fig. 7. All-*trans* retinoic acid (*AT-RA*) and 9-*cis* retinoic acid (*9C-RA*), and some typical synthetic retinoids: the retinoic acid receptor (RAR) α-selective *Am580* (Bernard et al. 1992), the RARβ/γ-selective *AGN190299* (Hembree et al. 1996), and the RARγ-selective *SR11254* (Dawson et al. 1995), in contrast to the retinoid X receptor-specific *SR11237* (Hembree et al. 1996)

like the 5,6,7,8-tetrahydro-5,5,8,8-tetramethyl-2-naphthalenyl group, which mimics the β-ionone moiety of RA, and on the other hand a carboxylate group which may be involved in a hydrophilic network similar to that of the RA-isomers (Klaholz et al. 1998).

RAR is an intriguing protein, as it binds the all-*trans* and the 9-*cis* isomers with almost the same affinity (Allenby et al. 1993; Allegretto et al. 1993), leading to comparable transactivation properties. This is in contrast to the expected shape of the unbound ligands, where AT-RA should be linear and 9C-RA quite bent. The promiscuous binding of 9C-RA to both RAR and RXR, but the lack of affinity of AT-RA to RXR could be due to two effects (Klaholz et al. 1998). First, the orientation of the 20-methyl group in AT-RA may lead to steric clashes with residues in the RXR ligand-binding site (for numbering see Fig. 7). And second, RXR most probably has a bent pocket, excluding the linear AT-RA but accomodating 9C-RA in a bent conformation. Consistently, conformationally restricted RXR-selective compounds exhibit a pronounced bend when compared to RAR-selective compounds, and often the 19-methyl group is magnified by additional groups such as in the ligand SR11237 (see Fig. 7). Antagonists appear to function by steric clash of bulky substituents with residues of the ligand-binding pocket, especially with H12, as recently confirmed by the structure of the ER LBD bound to raloxifene (Brzozowski et al. 1997). This disturbs the position of H12 and hinders the formation of the active receptor conformation, thereby preventing interaction with coactivators.

The question of selectivity stresses the need for further structural information. The present results already provide a structural basis for the development of more selective RAR ligands, hopefully less prone to side effects in medical application.

Acknowledgments. We thank J. P. Renaud for kindly providing Fig. 6 and for careful reading of the manuscript. We gratefully acknowledge the financial support of B. P. Klaholz by the Deutscher Akademischer Austauschdienst, Doktorandenstipendium aus Mitteln des dritten Hochschulsonderprogramms (HSPIII).

References

Allegretto EA, McClurg MR, Lazarchik SB, Clemm DL, Kerner SA, Elgort MG, Boehm MF, White SK, Pike JW, Heyman RA (1993) Transactivation properties of retinoic acid and retinoid X receptors in mammalian cells and yeast. Correlation with hormone binding and effects of metabolism [published erratum appears in J Biol Chem 1994 Mar 11; 269 (10):7834]. J Biol Chem 268:26625–26633

Allenby G, Bocquel MT, Saunders M, Kazmer S, Speck J, Rosenberger M, Lovey A, Kastner P, Grippo JF, Chambon P et al (1993) Retinoic acid receptors and retinoid X receptors: interactions with endogenous retinoic acids. Proc Natl Acad Sci USA 90:30–34

Bernard BA, Bernardon JM, Delescluse C, Martin B, Lenoir MC, Maignan J, Charpentier B, Pilgrim WR, Reichert U, Shroot B (1992) Identification of synthetic retinoids with selectivity for human nuclear retinoic acid receptor gamma. Biochem Biophys Res Commun 186:977–983

Bourguet W, Ruff M, Chambon P, Gronemeyer H, Moras D (1995) Crystal structure of the ligand-binding domain of the human nuclear receptor RXR-alpha. Nature 375:377–382

Brzozowski AM, Pike ACW, Dauter Z, Hubbard RE, Bonn T, Engström O, Öhman L, Greene GL, Gustafsson JA, Carlquist M (1997) Molecular basis of agonism and antagonism in the oestrogen receptor. Nature 389:753–758

Chambon P (1996) A decade of molecular biology of retinoic acid receptors. FASEB J 10:940–954

Chen JD, Evans RM (1995) A transcriptional co-repressor that interacts with nuclear hormone receptors. Nature 377:454–457

Chen JD, Umesono K, Evans RM (1996) SMRT isoforms mediate repression and anti-repression of nuclear receptor heterodimers. Proc Natl Acad Sci USA 93:7567–7571

Dawson MI, Chao WR, Pine P, Jong L, Hobbs PD, Rudd CK, Quick TC, Niles RM, Zhang XK, Lombardo A et al (1995) Correlation of retinoid binding affinity to retinoic acid receptor alpha with retinoid inhibition of growth of estrogen receptor-positive MCF-7 mammary carcinoma cells. Cancer Res 55:4446–4451

De Luca LM (1991) Retinoids and their receptors in differentiation, embryogenesis, and neoplasia. FASEB J 5:2924–2933

Driscoll JE, Seachord CL, Lupisella JA, Darveau RP, Reczek PR (1996) Ligand-induced conformational changes in the human retinoic acid receptor detected using monoclonal antibodies. J Biol Chem 271:22969–22975

Escriva H, Safi R, Hänni C, Langlois MC, Saumitou-Laprade P, Stehelin D, Capron A, Pierce R, Laudet V (1997) Ligand binding was aquired during evolution of nuclear receptors. Proc Natl Acad Sci USA 94:6803–6808

Evans SV (1993) Setor: hardware lighted three-dimensional solid model representations of macromolecules. J Mol Graphics 11:134–138

Giguere V (1994) Retinoic acid receptors and cellular retinoid binding proteins: complex interplay in retinoid signaling. Endocr Rev 15:61–79

Gronemeyer H, Moras D (1995) Nuclear receptors. How to finger DNA. Nature 375:190–191

Hembree JR, Agarwal C, Beard RL, Chandraratna RA, Eckert R (1996) Retinoid X receptor-specific retinoids inhibit the ability of retinoic acid receptor-specific retinoids to increase the level of insulin-like growth factor binding protein-3 in human ectocervical epithelial cells. Cancer Res 56:1794–1799

Horwitz KB, Jackson TA, Bain DL, Richer JK, Takimoto GS, Tung L (1996) Nuclear receptor coactivators and corepressors. Mol Endocrinol 10:1167–1177

Hörlein AJ, Naar AM, Heinzel T, Torchia J, Gloss B, Kurokawa R, Ryan A, Kamei Y, Soderstrom M, Glass CK et al (1995) Ligand-independent repres-

sion by the thyroid hormone receptor mediated by a nuclear receptor co-repressor. Nature 377:397–404
Kastner P, Mark M, Chambon P (1995) Nonsteroid nuclear receptors: what are genetic studies telling us about their role in real life? Cell 83:859–869
Keidel S, Le Motte P, Apfel C (1994) Different agonist- and antagonist-induced conformational changes in retinoic acid receptors analyzed by protease mapping. Mol Cell Biol 14:287–298
Klaholz BP, Renaud JP, Mitschler A, Zusi C, Chambon P, Gronemeyer H, Moras D (1998) Conformational adaptation of agonists to the human nuclear receptor RARγ. Nat Struct Biol 5:199–202
Kurokawa R, Di Renzo J, Boehm M, Sugarman J, Gloss B, Rosenfeld MG, Heyman RA, Glass CK (1994) Regulation of retinoid signalling by receptor polarity and allosteric control of ligand binding. Nature 371:528–531
Lee MS, Kliewer SA, Provencal J, Wright PE, Evans RM (1993) Structure of the retinoid X receptor alpha DNA binding domain: a helix required for homodimeric DNA binding. Science 260:1117–1121
Leid M (1994) Ligand-induced alteration of the protease sensitivity of retinoid X receptor alpha. J Biol Chem 269:14175–14181
Leng X, Tsai SY, O'Malley BW, Tsai MJ (1993) Ligand-dependent conformational changes in thyroid hormone and retinoic acid receptors are potentially enhanced by heterodimerization with retinoic X receptor. J Steroid Biochem Mol Biol 46:643–661
Leng X, Blanco J, Tsai SY, Ozato K, O'Malley BW, Tsai MJ (1995) Mouse retinoid X receptor contains a separable ligand-binding and transactivation domain in its E region. Mol Cell Biol 15:255–263
Levin AA, Sturzenbecker LJ, Kazmer S, Bosakowski T, Huselton C, Allenby G, Speck J, Kratzeisen C, Rosenberger M, Lovey A et al (1992) 9-cis retinoic acid stereoisomer binds and activates the nuclear receptor RXR alpha. Nature 355:359–361
Lotan R (1997) Retinoids and chemoprevention of aerodigestive tract cancers. Cancer Metastasis Rev 16:349–356
Luisi BF, Xu WX, Otwinowski Z, Freedman LP, Yamamoto KR, Sigler PB (1991) Crystallographic analysis of the interaction of the glucocorticoid receptor with DNA. Nature 352:497–505
Mader S, Chen JY, Chen Z, White J, Chambon P, Gronemeyer H (1993) The patterns of binding of RAR, RXR and TR homo- and heterodimers to direct repeats are dictated by the binding specificites of the DNA binding domains. EMBO J 12:5029–5041
Mangelsdorf DJ, Evans RM (1995) The RXR heterodimers and orphan receptors. Cell 83:841–850

Mangelsdorf DJ, Thummel C, Beato M, Herrlich P, Schutz G, Umesono K, Blumberg B, Kastner P, Mark M, Chambon P et al (1995) The nuclear receptor superfamily: the second decade. Cell 83:835–839

McInerney EM, Tsai MJ, O'Malley BW, Katzenellenbogen BS (1996) Analysis of estrogen receptor transcriptional enhancement by a nuclear hormone receptor coactivator. Proc Natl Acad Sci USA 93:10069–10073

Napoli JL (1996) Retinoic acid biosynthesis and metabolism. FASEB J 10:993–1001

Nicholls A, Sharp KA, Honig B (1991) Protein folding and association: insights from the interfacial and thermodynamic properties of hydrocarbons. Proteins 11:281–286

Niles RM (1995) Use of vitamins A and D in chemoprevention and therapy of cancer: control of nuclear receptor expression and function. Vitamins, cancer and receptors. Adv Exp Med Biol 375:1–15

Oñate SA, Tsai SY, Tsai MJ, O'Malley BW (1995) Sequence and characterization of a coactivator for the steroid hormone receptor superfamily. Science 270:1354–1357

Rastinejad F, Perlmann T, Evans RM, Sigler PB (1995) Structural determinants of nuclear receptor assembly on DNA direct repeats. Nature 375:203–211

Renaud JP, Rochel N, Ruff M, Vivat V, Chambon P, Gronemeyer H, Moras D (1995) Crystal structure of the RAR-gamma ligand-binding domain bound to all-trans retinoic acid. Nature 378:681–689

Rochette-Egly C, Adam S, Rossignol M, Egly JM, Chambon P (1997) Stimulation of RAR-alpha activation function AF-1 through binding to the general transcription factor TFIIH and phosphorylation by cdk7. Cell 90:97–107

Schwabe JWR, Chapman L, Finch JT, Rhodes D, Neuhaus D (1993) DNA recognition by the oestrogen receptor: from solution to the crystal. Structure 1:187–204

Tallman MS, Wiernik PH (1992) Retinoids in cancer treatment. J Clin Pharmacol 32:868–888

Voegel JJ, Heine MJ, Zechel C, Chambon P, Gronemeyer H (1996) TIF2, a 160 kDa transcriptional mediator for the ligand-dependent activation function AF-2 of nuclear receptors. EMBO J 15:3667–3675

Wagner RL, Apriletti JW, McGrath ME, West BL, Baxter JD, Fletterick RJ (1995) A structural role for hormone in the thyroid hormone receptor. Nature 378:690–697

Wurtz JM, Bourguet W, Renaud JP, Vivat V, Chambon P, Moras D, Gronemeyer H (1996) A canonical structure for the ligand-binding domain of nuclear receptors. Nat Struct Biol 3:87–94

Ylikomi T, Bocquel MT, Berry M, Gronemeyer H, Chambon P (1992) Cooperation of proto-signals for nuclear accumulation of estrogen and progesterone receptors. EMBO J 11:3681–3694

Zechel C, Shen XQ, Chambon P, Gronemeyer H (1994a) Dimerization interfaces formed between the DNA binding domains determine the cooperative binding of RXR/RAR and RXR/TR heterodimers to DR5 and DR4 elements. EMBO J 13:1414–1424

Zechel C, Shen XQ, Chen JY, Chen ZP, Chambon P, Gronemeyer H (1994b) The dimerization interfaces formed between the DNA binding domains of RXR, RAR and TR determine the binding specificity and polarity of the full-length receptors to direct repeats. EMBO J 13:1425–1433

9 Characterization of Human Estrogen Receptor β

E. Enmark and J.-Å. Gustafsson

9.1 Introduction

Estrogens influence the growth, differentiation, and function of many target tissues, including tissues of the male and female reproductive systems such as mammary gland, uterus, vagina, ovary, testis, epididymis, and prostate (Clark et al. 1992). Estrogens also play important roles in bone maintenance (Turner et al. 1994) and in the cardiovascular system where they have been shown to have cardioprotective effects (Farhat et al. 1996), and in the central nervous system (Mccarthy and Pfaus 1996). Natural and synthetic antagonists are used in a number of clinical applications such as, for example, breast cancer therapy, treatment and prevention of osteoporosis, and in estrogen replacement therapy of postmenopausal women. Estrogens are mainly produced in the ovaries and testes. The three main forms of estrogens are 17β-estradiol (biologically the most potent), estriol, and estrone. They diffuse in and out of cells, but are retained with high affinity and specificity in target cells by an intranuclear binding protein, termed the estrogen receptor (ER). This receptor regulates the expression of specific target genes by binding to target sequences called response elements in the promoter region of the respective target genes (Gronemeyer and Laudet 1995).

The nuclear receptor superfamily includes, in addition to the steroid receptors, thyroid, retinoic acid, and vitamin D3 receptors. In addition, a large number of "orphan" receptors have been identified, for which so far no ligand has been identified. This protein family has, as of today, approximately 70 known members, about two thirds of which are orphan receptors (Enmark and Gustafsson 1996).

9.2 The Estrogen Receptor

Analysis of ER and other steroid receptors shows that they can be subdivided into several functional domains (Beato et al. 1995). The N-terminal A/B domain is highly variable in sequence and length, and usually contains a transactivation function, which activates target genes by interacting with components of the core transcriptional machinery. The DNA-binding domain, also referred to as the C-region, contains two zinc fingers, which are involved in specific DNA binding and receptor dimerization. The hinge (D) domain contributes flexibility to the DNA- versus the ligand-binding domain, and has also in some cases been shown to influence the DNA-binding properties of individual receptors. The ligand-binding domain (sometimes called the "E" domain) is relatively large and harbors regions important for ligand binding, receptor dimerization, nuclear localization, and interactions with transcriptional coactivators and corepressors. The C-terminal extension domain, finally, has been shown to contribute to the transactivation capacity of the receptor, but its other functions, if any, are to a large extent unknown.

One ER was cloned in 1986 from the uterus (Green et al. 1986). The dogma, which prevailed soon thereafter, of the existence of a single ER gene was difficult to reconcile with the striking tissue-specific differences in the action of synthetic estrogens and antiestrogens.

For many other members of the nuclear receptor superfamily multiple receptor subtypes have been identified, for instance in the case of the thyroid hormone and retinoic acid receptors, as well as for many of the orphan receptors (Gronemeyer and Laudet 1995).

Furthermore, when an ERα knock-out mouse was developed, specific estrogen binding was still observed in some tissues (Lubahn et al. 1993). Surprisingly, in these ER knock-out (ERKO) mice, apart from large abnormalities in the reproductive organs, the gene deletion had little or no effect on, for example, bone stability or on the cardiovascular system (Lubahn et al. 1993). In retrospect, this suggested the existence of a second receptor for estrogen.

9.3 Estrogen Receptor β: Basic Properties

The cloning and initial characterization of rat ERβ was described in 1996 (Kuiper et al. 1996a). ERβ is highly homologous to the previously identified estrogen receptor (consequently called ERα), particularly in the DNA-binding domain (95% amino acid identity) and in the ligand-binding domain (55% amino acid identity) (Fig. 1). Ligand-binding experiments revealed high affinity, high specificity, and capacity binding of estradiol by ERβ, with a K_d similar to that of ERα, 0.6 nM.

The ERβ protein is capable of stimulating transcription of an ER target gene in a manner similar to ERα, using synthetic estrogen response elements in front of reporter genes.

Some natural and synthetic ligands show differences in binding affinity to ERα and ERβ, although many ligands bind with similar affinity. This suggests that it should be possible to develop ligands specific to either of the two subtypes.

In the rat, ERα shows highest expression in uterus, testis, pituitary, ovary, kidney, epididymis and adrenal, and ERβ shows the highest

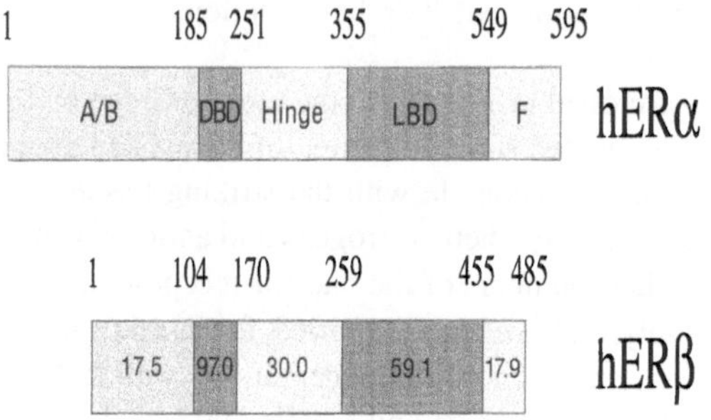

Fig. 1. Comparison between the human estrogen receptor α (*hER*α) and human ERβ (*hER*β). Numbers above each receptor represents amino acid numbers, whereas numbers inside the respective boxes represent percentage of amino acid identity

expression in prostate, ovary, lung, bladder, brain and epididymis (Kuiper et al. 1996b).

We and others have reported that ERα and ERβ are able to form heterodimers, both in solution and when bound to DNA; heterodimer formation is greatly enhanced by the presence of ligand (Pettersson et al. 1997).

9.4 The ERβ Gene

The exon/intron organization of both the mouse and human ERβ genes has been determined (Enmark et al. 1997). The translated exons of the mouse ERβ gene span approximately 40 kb. The analysis of the human ERα gene has shown that it is a very large gene, with the translated exons spanning more than 140 kb (Ponglikitmongkol et al. 1988). The ER genes from fish, however, are considerably smaller spanning approximately 30–40 kb (LeRoux et al. 1993). Our data show that the size of the ERβ gene is similar to that of the fish ERs. For other paralogous genes in the nuclear receptor superfamily, the gene size has been shown to vary at least to the same extent as for the ER subtypes. The peroxisome proliferator-activated receptor (PPAR) genes, for example, differ in size from between 30 and 105 kb in the mouse. It has been speculated that the size of the introns might influence the transcriptional efficiency of a gene, particularly in situations of rapid cell division (O'Farrell 1992), a notion that might well be true for prokaryotes and lower eukaryotes, which are organisms with a short generation time. The relevance of this phenomenon for mammalian genes has never been verified.

The mouse ERβ gene has recently been mapped to mouse chromosome 12 using interspecific backcross analysis (Tremblay et al. 1997). This represents a chromosomal region homologous to 14q22–24, where the human ERβ gene is localized (Enmark et al. 1997). Since the human ERα gene is located on the long arm of chromosome 6, this definitely excludes the possibility of differential splicing to explain the formation of the ERβ subtype. 14q22–24 is close to a recently identified gene associated with early onset of Alzheimer's disease (Sherrington et al. 1995) and this region is also frequently involved in rearrangements in human uterine leiomyoma and in neoplasms of the kidney. A more detailed mapping of this chromosomal region, as well as studies on patient material, will in time tell whether this chromosomal localization of ERβ has any relevance with reference to the diseases mentioned.

All exon/intron boundaries are well conserved in the ERβ gene as compared to the human ERα gene (Fig. 2). Notably, the only difference observed in the genomic organization of the ER genes is the intron present in the middle of the D domain of the ER isolated from rainbow trout and *Oreochromis aureus* which is absent from both ERα and ERβ. Interestingly, sequence comparison of all known ERs shows that these receptors seem to form three groups, where the receptors cloned from fish constitute a separate subgroup. The exception in the fish subgroup is the ER cloned from japanese eel, which actually represents an ERβ homologue (Todo et al. 1996).

The question of whether the third subgroup represents an "ERγ", or whether in general there are further subtypes of ERs is important. Mice with double knock-outs of both the ERα and ERβ genes will obviously be interesting to investigate in this respect. If specific estrogen binding can still be observed, this would indicate that further ER subtypes might exist.

All subtypes of the ER show very complex patterns at the level of transcriptional control. ERα has been shown in different species to have at least two or three separate promoters with different but overlapping

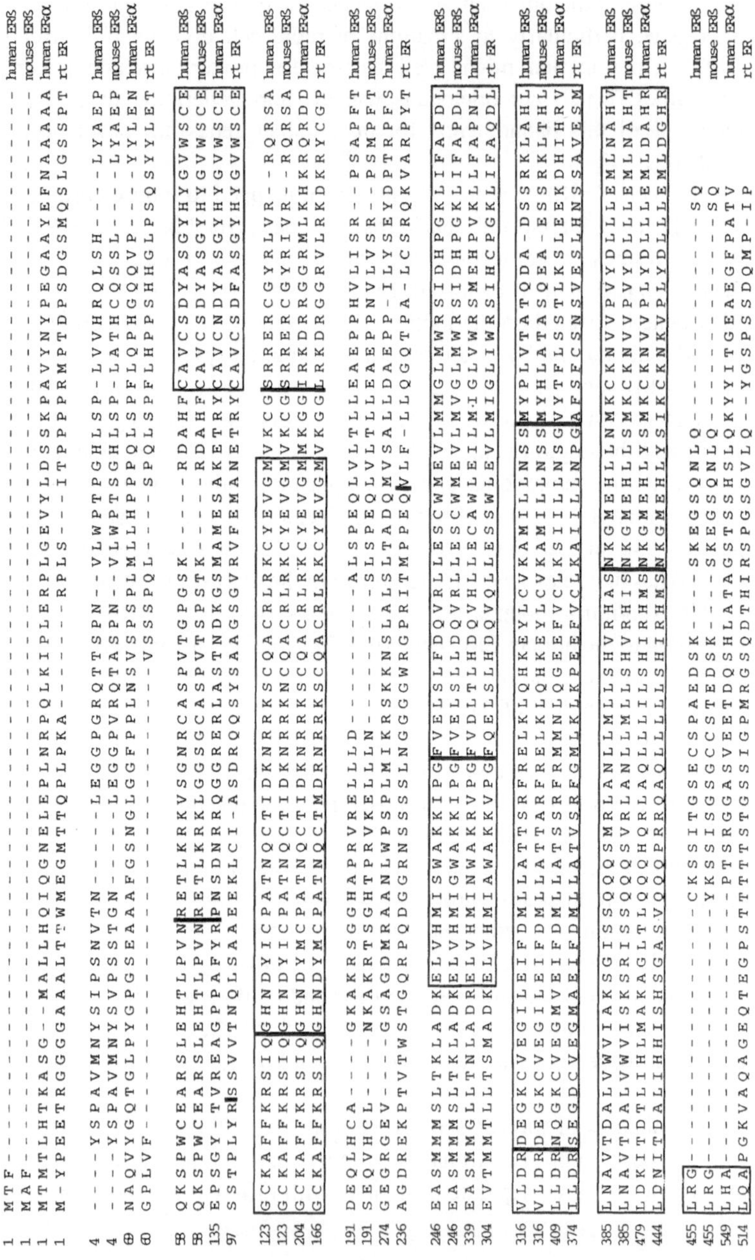
human ERβ
mouse ERβ
human ERα
rt ER

tissue distribution (Grandien et al. 1995). In the *O. aureus* ER, one of the ERs isolated from fish, two alternative polyadenylation sites located approximately 300 bp apart, have been found, in addition to two different transcription start sites (Tan et al. 1996). Both mouse and human ERβ give several bands of very different sizes in Northern blot experiments (Enmark et al. 1997; Tremblay et al. 1997), possibly indicating that also the ERβ gene is characterized by multiple promoters and/or polyadenylation signals.

In addition, several groups have recently reported on isoforms of ERβ also at the protein level, isoforms which differ either in the amino terminus or in the ligand-binding domain (Chu and Fuller 1997). No biological function for these isoforms has as yet been defined. In in vitro experiments, at least one of the isoforms in the ligand-binding domain binds to estradiol with a much lower affinity than the canonical ERβ receptor. In the nuclear receptor superfamily such isoforms are relatively common, particularly in the N-terminal domain, and have been most carefully studied in the retinoic acid receptors (Gronemeyer and Laudet 1995). Before attributing too much weight to the ERβ isoforms it is mandatory that their presence should be assessed not only at the mRNA but also at the protein level. Previous work on ERα isoforms has not yet proven conclusively that these receptor variants are essential in the estrogen mechanism of action.

9.5 Human ERβ

Over the past two years ERβ has also been cloned from human (Mosselman et al. 1996; Enmark et al. 1997), mouse (Tremblay et al. 1997) and marmoset monkey (*Callithrix jacchus*) (J. Gaughan, unpublished, Genbank no. Y09372). Human ERβ shows approximately 89% identity to rat ERβ, 88% identity to mouse ERβ, and 47% identity to human ERα, in its translated portion, a degree of similarity well in concordance with

---

Fig. 2. Comparison of intron positions in the human and mouse estrogen receptor β (*human/mouse ER*β), human estrogen receptor α (*human ER*α) and rainbow trout ER (*rt ER*). The alignment was created using the Clustal alignment tool and the MegAlign program of Lasergene

that observed for species homologues of other nuclear receptors. It can be noted, however, that the degree of homology between ERα and ERβ is unusually low for two receptor subtypes, particularly in the ligand-binding domain.

9.6 Tissue Distribution of Human ERβ

The distribution of ERβ in human tissues has been studied to various extents in many groups using different methods, i.e., reverse transcriptase polymerase chain reaction assay, in situ hybridization, and immunohistochemistry. These studies have shown that ERβ is highly expressed in several human organs, both in reproductive tissues and in some tissues probably erroneously considered as "non-targets" for estrogen (Fig. 3).

9.7 "Non-Target" Tissues

Some examples of this category of tissue are found in the gastrointestinal tract where ERβ mRNA is highly expressed in the mucosa of the stomach, duodenum, colon, and rectum (Enmark et al. 1997). We have shown that ERβ has a relatively high affinity for several plant-derived substances with estrogenic activity, considerably higher than ERα (Kuiper et al. 1996). It is possible that ERβ expressed in the gastrointestinal tract is exposed to these compounds via diet. For several years now it has been suggested that estrogens may protect against colon cancer (Newcomb and Storer 1995). Similar claims have also been made for diets containing soy protein, a product rich in phytoestrogens (Goldin et al. 1995). Estrogens have furthermore been shown to affect calcium uptake in the intestine through a poorly understood mechanism (Arjmandi et al. 1994), and perhaps ERβ may mediate some of these effects.

In the kidney, high expression of ERβ is seen in convoluted tubules in the cortex. Also the transitional epithelium in renal pelvis expresses ERβ (Enmark et al. 1997). In the hamster, estrogen-induced kidney cancer is a well known tumor model system, where it now seems possible that ERβ might be involved.

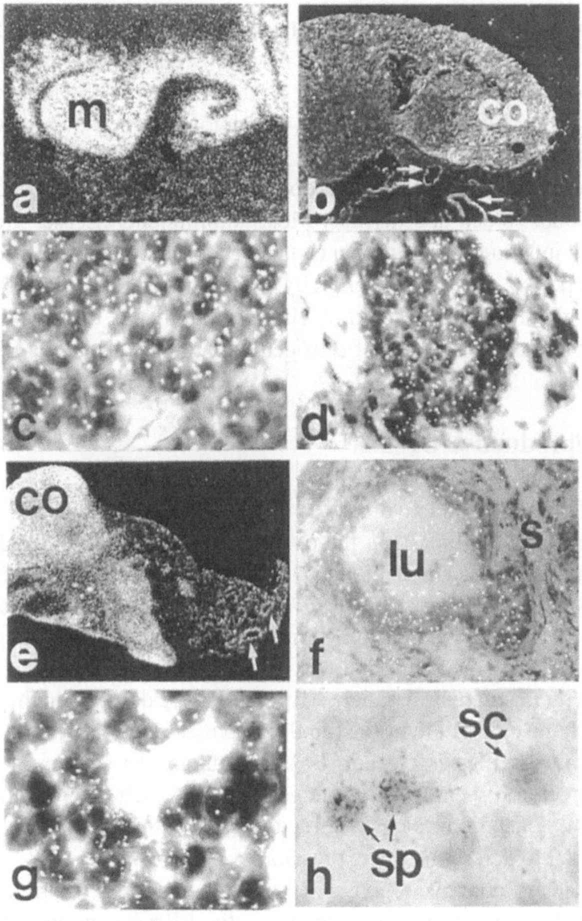

Fig. 3a–h. Expression of estrogen receptor β (ERβ) in human, studied by in situ hybridization. The general impression is that expression of ERβ is detected in a larger variety of tissues. Some examples are shown. **a** Mucosa cells (*m*) of the duodenum express high levels of ERβ mRNA. **b** In the kidney ERβ is found in the cortex (*co*) and in blood vessels (marked by *arrows*). **c** Most of the cells of the lung express ERβ. **d** Breast tumor. We have found ERβ in several tumors of the breast, both in ERα-positive and -negative samples. **e** In ovaries ERβ is found in the cortex (*co*), in blood vessels (marked by *arrows*), and also in human granulosa cells (not shown). **f** In prostate a signal is present in the epithelium of secretory alveoli (*lu*), while surrounding stroma (*s*) does not have detectable levels of ERβ. **g** A large portion of the lymphocytes in lymph node express ERβ. **h** In the testis ERβ mRNA is found in spermatocytes (*sp*), but not in Leydig cells or Sertoli cells (*sc*)

In lung, ERβ is expressed both in the parenchyma and in the blood vessels (Enmark et al. 1997).

In lymph nodes and the thymus a large portion of the lymphocytes express ERβ (Enmark et al. 1997). It has long been known that estrogens have important effects on the immune system. In pregnancy the immune system is significantly downregulated, leading to decreased size of both spleen and thymus. Most autoimmune diseases are more common in women than in men (Lahita 1996). Estrogens have recently been reported to also specifically inhibit thymocyte development (Rijhsinghani et al. 1997). An exciting possibility is that some of the immunomodulatory effects of estrogen might be mediated via ERβ.

9.8 Reproductive Organs

In the female reproductive tissues, ERβ is found in the ovary, uterus, endometrium, and breast. In the ovary, the receptor is localized to the stroma of the cortex and in blood vessels of the medulla, as well as to the granulosa cells (Enmark et al. 1997; Byers et al. 1997). The granulosa cells apparently contain only ERβ mRNA. ERβ is thus likely to play an important role in the regulation of follicular growth and oocyte development. With in situ hybridization, ERβ can also be detected in the uterus. In breast, the epithelium of the tubules expresses ERβ, and also in some breast cancers ERβ is present. In breast tumors ERα expression and ERβ expression seem to vary independently of each other (Dotzlaw et al. 1997). For the future characterization of breast tumors it might thus be relevant to determine the expression of both ERs.

In male reproductive organs, ERβ is expressed in the testis and in the prostate (Enmark et al. 1997). In the testis, ERα has previously been reported to be expressed in the Leydig cells of the testis (Fischer et al. 1997), where we find no ERβ signal. In contrast, ERβ is expressed in developing spermatids, where ERα is absent.

During recent years, there has been an intensive debate concerning alleged effects of different xenobiotics on the reproductive ability of animals, particularly in fish and man. A class of compounds called "environmental estrogens", including, e.g., polychlorinated biphenyls, has been in particular focus (Jobling et al. 1995). We have shown that both ERα and ERβ may bind at least some of these compounds (Kuiper et al. 1996b). Although the affinity is relatively low, ERβ binds the xenoestrogens methoxychlor and bisphenol A with considerably higher affinity than ERα. As human ERβ is expressed in the developing spermatocytes of the testis, it is tempting to speculate that some of the claimed effects of environmental estrogens on fertility might be mediated via ERβ.

In the prostate, ERβ is expressed in the epithelial cells of the secretory alveoli. Although expression in the human prostate seems to be lower than in the rat prostate this might still be relevant for the development and progression of prostate cancer and/or benign prostatic hyperplasia. Tumors in the prostate are initially usually dependent on testosterone and orchidectomy and anti-androgens have been used in the treatment of prostate cancer. Also estrogens are sometimes effective as treatment, unfortunately, however, leading to an increased risk of cardiovascular complications. Prostatic hyperplasia is an extremely common condition among elderly men, and is a disease which is possibly linked to a shift in the balance between androgens and estrogens in the blood.

9.9 Bone

Surprisingly, studies of ERKO mice (see Sect. 9.2) show that gene deletion has little or no effect on bone stability. Studies by our own and other groups have indeed shown that ERβ is expressed in osteoblasts of growing bones in the rat. ERβ is also detected in osteoblastic cell lines, pointing to a possible role of ERβ in bone formation (Arts et al. 1997; Onoe et al. 1997). The relevance or these observations in the human is, however, unclear since the only known human patient lacking functional ERα displays severe osteoporosis (Smith et al. 1994).

9.10 The Cardiovascular System

It was recently shown that in ERKO mice the atheroprotective effect of estrogen is unchanged, using a carotid arterial injury model (Iafrati et al. 1997). The authors conclude that the protective effect of estrogen is independent of ERα. We have found that ERβ, but not ERα, is expressed in human umbilical vein endothelial cells, a finding well in line with these observations in mice.

9.11 The Central Nervous System

Effects of estrogens on the central nervous system have been known for many years. It has for example been claimed that estrogen replacement therapy reduces the risk of Alzheimer's disease in women, as well as improving this condition, at least in some patients (Tang et al. 1996). Both ERα and ERβ are present in the brain, but ERβ seems to be the main estrogen receptor in the brain in, for example, the hypothalamus, the hippocampus, and in several limbic regions (Li et al. 1997).

9.12 Two Receptors: Multiple Possiblilities

The recent discovery that an additional ER subtype (ERβ) is present in various rat, mouse, and human tissues has significantly advanced our understanding of the mechanisms underlying estrogen signaling. It suggests the existence of two previously unrecognized pathways of estrogen signaling: via ERβ in cells exclusively expressing this subtype and via ERα/ERβ heterodimers in cells expressing both ER subtypes (Fig. 4). It cannot be excluded that ERβ homodimers interact with novel response elements, apart from the known estrogen response elements (EREs). It must also be kept in mind that only a few of the estrogen-regulated genes contain classical EREs. Both ERα and ERβ may activate genes via binding to activating protein 1 (AP-1) sites, and it has

Fig. 4. Three alternative signalling pathways for estrogen. The existence of two estrogen receptor (ER) subtypes (*ERα*, *ERβ*) that may heterodimerize creates three possible modes of gene activation. In cells which only express one subtype, homodimers of ERα or ERβ interact with response elements in the respective promotors of the target genes. In cells that express both receptors, heterodimers between ERα and ERβ may also be formed to various extents, depending on the relative levels of the two receptors. It is still an open question whether specific target genes exist that are activated only by one ER homodimer or only by the receptor heterodimer. *LBD*, ligand-binding domain; *DBD*, DNA-binding domain; *RE*, response element; *ERE*, estrogen response element ▶

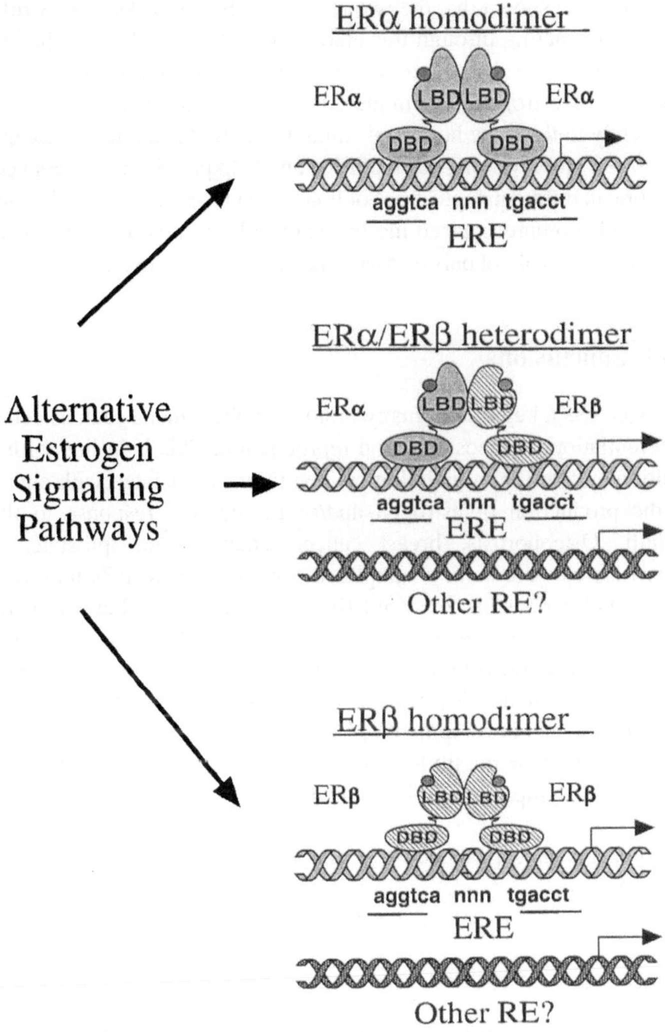

Fig. 4. Legend see. p.172

recently been shown that the two receptor subtypes indeed show differences when acting through this element (Paech et al. 1997). The existence of two ER subtypes greatly expands the physiological regulatory potential of estrogenic hormones. Different target cells may respond differently to the same hormonal stimulus due to the alternative composition of receptors. Varying ratios of ERα and ERβ proteins in different cells, resulting in different populations of homo- and heterodimers, could constitute a hitherto unrecognized mechanism involved in tissue- and cell type-specific actions of not only estrogens but also of antiestrogens.

9.13 Conclusions

Estrogen has a key role in many biological phenomena such as cellular differentiation, homeostasis, and reproduction. This is in line with the multitude of different pathological conditions associated with changes in the production of estrogen and/or the cellular response to these stimuli. Osteoporosis, breast cancer, endometrium, prostate, and atherosclerosis are some examples of diseases where ERs may be involved. Our recent discovery of ERβ shows that the mechanisms behind the effects of estrogen are far more complicated than previously assumed, and provides unique opportunities to gain a better understanding of these phenomena. ERβ is expressed in many important target tissues for estrogen, and development of ERα- and ERβ-specific ligands opens up interesting new possibilities for the treatment of, for example, postmenopausal symptoms and breast cancer.

Acknowledgements. This study was supported by a grant from the Swedish Cancer Society.

References

Arjmandi BH, Hollis BW, Kalu DN (1994) In vivo effect of 17β-estradiol on intestinal calcium absorption in rats. Bone Miner 26:181–189

Arts J, Kuiper GG, Janssen JM, Gustafsson J-Å, Löwik CW, Pols HA, van Leeuwen JP (1997) Differential expression of estrogen receptors α and β mRNA during differentiation of human osteoblast SV-HFO cells. Endorcrinology 138:5067–5070

Beato M, Herrlich P, Schutz G (1995) Steroid hormone receptors: many actors in search of a plot. Cell 83:851–857

Byers M, Kuiper GGJM, Gustafsson J-Å, Park-Sarge O-K (1997) Estrogen receptor β mRNA expression in rat ovary: downregulation by gonadotropins. Mol Endocrinol 11:172–182

Chu S, Fuller P (1997) Identification of a splice variant of the rat estrogen receptor beta gene. Mol Cell Endocrinol 132:195–199

Clark JH, Schrader WT, O'Malley BW (1992) Mechanisms of action of the steroid hormone. In: Wilson JD, Foster DW (eds) Textbook of endocrinology. Saunders, New York, pp 35–90

Dotzlaw H, Leygue E, Watson PH, Murphy LC (1997) Expression of estrogen receptor β in human breast tumors. J Clin Endocrinol Metab 82:2371–2374

Enmark E, Gustafsson J-Å (1996) Orphan nuclear receptors – the first eight years. Mol Endocrinol 10:1293–1307

Enmark E, Pelto-Huikko M, Grandien K, Fried G, Lagerkrantz S, Lagerkrantz J, Nordenskjöld M, Gustafsson J-Å (1997) Human estrogen receptor β – gene structure, chromosomal localisation and expression pattern. J Clin Endocrinol Metab 82:4258–4265

Farhat MY, Lavigne MC, Ramwell PW (1996) The vascular protective effects of estrogen. FASEB J 10:615–624

Fischer JS, Millar MR, Majdic G, Saunders PTK, Fraser HM, Sharpe RM (1997) Immunolocalisation of oestrogen receptor-alpha within the testis and excurrent ducts of the rat and marmoset monkey from perinatal life to adulthood. J Endocrinol 153:485–495

Goldin BR, Gorbach SL, Hockerstedt KA, Watanabe S, Hamalainen EK, Markkanen MH, Makela TH, Wahala KT, Adlercreutz CH. (1995) Soybean phytoestrogen intake and cancer risk. J Nutr 125:757S–770S

Grandien K, Backdahl M, Ljunggren O, Gustafsson J-Å, Berkenstam A (1995) Estrogen target tissue determines alternative promoter utilization of the human estrogen receptor gene in osteoblasts and tumor cell lines. Endocrinology 136:2223–2229

Green S, Walter P, Kumar V, Krust A, Bornet J-M, Argos P, Chambon P (1986) Human oestrogen receptor cDNA: sequence, expression and homology to v-erbA. Nature 320:134–139

Gronemeyer H, Laudet V (1995) Transcription factors 3: nuclear receptors. Protein Profile 2:1173–1308

Iafrati MD, Karas RH, Aronovitz M, Kim S, Sullivan TR, Lubahn DB, O'Donnell TF Jr, Korach KS, Mendelsohn ME (1997) Estrogen inhibits the vascular injury response in estrogen receptor α-deficient mice. Nat Med 3:545–548

Jobling S, Reynolds T, White R, Parker MG, Sumpter JP (1995) A variety of environmentally persistent chemicals, including some phtalate plasticizers, are weakly estrogenic. Environ Health Perspect 103:582–587

Kuiper G, Enmark E, Pelto-Huikko M, Nilsson S, Gustafsson J-Å (1996a) Cloning of a novel estrogen receptor expressed in rat prostate and ovary. Proc Natl Acad Sci USA 93:5925–5930

Kuiper GGJM, Carlsson B, Grandien K, Enmark E, Häggblad J, Nilsson S, Gustafsson J-Å (1996b) Comparison of the ligand binding specificity and transcript tissue distribution of estrogen receptors α and β. Endocrinology 138:863–870

Lahita RG (1996) The connective tissue diseases and the overall influence of gender. Int J Fertil Menopausal Stud 41:156–165

LeRoux M-G, Thézé N, Wolff J, Le Pennec JP (1993) Organization of a rainbow trout estrogen receptor gene. Biochim Biophys Acta 1172:226–230

Li X, Schwartz PE, Rissman EF (1997) Distribution of estrogen receptor-beta-like immunoreactivity in rat forebrain. Neuroendocrinology 66:63–67

Lubahn DB, Moyer, JS, Golding TS, Couse JF, Korach KS, Smithies O (1993) Alteration of reproductive function but not prenatal sexual development after insertional disruption of the mouse estrogen receptor gene. Proc Natl Acad Sci USA 90:11162–11166

Mccarthy MM, Pfaus JG (1996) Steroid modulation of neurotransmitter functin to alter female reproductive behavior. Trends Endocrinol Metabol 7:327–333

Mosselman S, Pohlman J, Dijkema R (1996) ERβ: identification and characterisation of a novel human estrogen receptor. FEBS Lett 392:49–53

Newcomb PA, Storer BE (1995) Postmenopausal hormone use and risk of large-bowel cancer. J Natl Cancer Inst 87:1067–1071

O'Farrell PH (1992) Big genes and little genes and deadlines for transcription. Nature 359:366–367

Onoe Y, Miyaura C, Ohta H, Nozawa S, Suda T (1997) Expression of estrogen receptor β in rat bone. Endocrinology 138:4509–4512

Paech K, Webb P, Kuiper GGJM, Nilsson S, Gustafsson J-Å, Kushner PJ, Scanlan TS (1997) Differential ligand activation of estrogen receptors ERα and ERβ at AP1 sites. Science 277:1508–1510

Pettersson K, Grandien K, Kuiper GGJM, Gustafsson J-Å (1997) Mouse estrogen receptor beta forms estrogen response element-binding heterodimers with estrogen receptor alpha. Mol Endocrinol 11:1486–1496

Ponglikitmongkol M, Green S, Chambon P (1988) Genomic organisation of the human oestrogen receptor gene. EMBO J 7:3385–3388

Rijhsinghani A, Bhatia SK, Kantmneni L, Schlueter A, Waldschmidt TJ (1997) Estrogen inhibits fetal thymocyte development in vitro. Am J Reprod Immunol 37:384–390

Sherrington R, Rogaev EI, Liang Y, Rogaeva EA, Levesque G, Ikeda M, Chi H, Lin C, Li G, Holman K et al (1995) Cloning of a gene bearing missense mutations in early onset familiar Alzheimer's disease. Nature 375:754–760

Smith EP, Boyd J, Frank GR, Takahashi H, Cohen RM, Specker B, Williams TC, Lubahn DB, Korach KS (1994) Estrogen resistance caused by a mutation in the estrogen.receptor gene in a man. N Engl J Med 331:1088–1089

Tan NS, Lam TJ, Ding JL (1996) The first contiguous estrogen receptor gene from a fish, Oreochromis aureus: evidence for multiple transcripts. Mol Cell Endocrinol 120:177–192

Tang M-X, Jacobs D, Stern Y, Schofield P, Gurland B, Andrews H, Mayeux R (1996) Effect of oestrogen during menopause on risk and age at onset of Alzheimers disease. Lancet 348:429–432

Todo T, Adachi S, Yamauchi K (1996) Molecular cloning and characterization of Japanese eel estrogen receptor cDNA. Mol Cell Endocrinol 119:37–45

Tremblay GB, Tremblay A, Copeland NG, Gilbert DJ, Jenkins NA, Labrie F, Giguère V (1997) Cloning, chromosomal localization, and functional analysis of the murine estrogen receptor β. Mol Endocrinol 11:353–365

Turner RT, Riggs BL, Spelsberg TC (1994) Skeletal effects of estrogens. Endocr Rev 15:275–300

10 Functional and Pharmacological Analysis of the A and B Forms of the Human Progesterone Receptor

P. Giangrande, G. Pollio, and D.P. McDonnell

10.1 Introduction

The steroid hormone progesterone is a key regulator of the processes involved in the development and maintenance of reproductive function (Clark and Peck 1979; Clarke and Sutherland 1990). In addition, however, the efficacy of antiprogestins as treatments for brain meningiomas, breast cancer, uterine fibroids, and endometriosis have implicated progesterone in the pathology of these diseases (Poisson et al. 1983;

Colletta et al. 1991; Kettel et al. 1991; Horwitz 1992; Lundgren 1992; Brandon et al. 1993; Carroll et al. 1993). The mechanism by which progesterone manifests biological activity in target tissues is similar to that of other members of the steroid hormone receptor superfamily (McDonnell 1995). In mammals progesterone is transported in the blood throughout the body and is capable of diffusing across all cell membranes; however, it exerts biological activity only in those cells which express a specific high affinity nuclear progesterone receptor (PR).

In the absence of progesterone, PR resides in an inactive state within the nuclei of target cells associated with a large multi-component heat-shock protein complex (McDonnell 1995). Upon hormone binding, however, the receptor undergoes a conformational change, an event which leads to the dissociation of the receptor from the inhibitory heat-shock protein complex and the subsequent formation of stable homodimers (Vegeto et al. 1992). In this form the receptor is capable of interacting with specific high affinity DNA sequences (progesterone response elements; PREs) located within the regulatory regions of target gene promoters and is capable of positively or negatively regulating target gene transcription. Interestingly, this latter step in the signal transduction pathway appears to be the most complex and may not occur in the same manner in all cells and on all promoters. In some instances, it is now considered that the receptor can directly contact components of the general transcriptional machinery and that, in doing so, it can stabilize the transcription initiation complex (Jacq et al. 1994; Mengus et al. 1995; Schwerk et al. 1995). However, in addition to this mechanism, it is clear that the activated receptor can interact with the general transcription machinery in an indirect manner, through any of a number of intermediary cofactors (Halachmi et al. 1994; Cavaillès et al. 1995; Le Douarin et al. 1995; Oñate et al. 1995; Hanstein et al. 1996; Hong et al. 1996; Voegel et al. 1996). Of course these different pathways are not mutually exclusive but are differentially utilized in different cells. Overlaying this tremendous complexity is the fact that the PR, in most mammalian species, exists within target cells in either of two forms, PR-A or PR-B (Lessey et al. 1983). This complexity, however, provides an explanation for the cell-selective activities which are manifest by different PR ligands, and suggests further that it may be possible to develop pharmaceuticals that function as progestins or antiprogestins in a tissue-selective manner. This chapter specifically focuses on the role of the A and B forms of the PR in PR pharmacology and considers their utility as drug targets.

10.2 Two Different Forms of the Human Progesterone Receptor Exist in Target Cells

In humans, the two PR isoforms A and B arise from unique mRNA transcripts produced from a single gene by alternate transcription initiation (Kastner et al. 1990; Gronemeyer et al. 1991). Interestingly, under most circumstances hPR-A and hPR-B are co-expressed in target cells in approximately equimolar amounts (Lessey et al. 1983). Thus, as a consequence of the process of dimerization, the activated receptor can exist in the cell in either of three states, A:A, A:B, and B:B (DeMarzo et al. 1991, 1992). Because, as will be discussed in Sect. 10.3, the A and B forms of PR are not functionally equivalent, it is implied that the three dimeric states are also not functionally identical. Thus, the relative expression of the two isoforms is likely to be important in regulating cellular responsiveness to progestins and antiprogestins.

The initial characterization of the A and B forms of hPR indicated that they had equivalent affinities for target DNA and exhibited indistinguishable ligand-binding affinities and specificitics (Lessey et al. 1983; Christensen et al. 1991). It was not apparent from these results, however, whether or not these PR isoforms were functionally different. In fact, it was suggested by some that the smaller A form of PR was derived in an artifactual manner from the B form as a result of proteolysis. However, the ontogeny of these receptors has now been elucidated for both chicken and human receptors. In the chicken, both receptors are derived from the same mRNA by alternate initiation of translation (Conneely et al. 1987). In humans, however, it appears as if both isoforms are derived from unique mRNAs which are produced from different promoters within the same gene (Kastner et al. 1990). These results, indicating that the cell has evolved complex regulatory mechanisms to regulate PR-A and PR-B expression, strongly suggested that both forms were important for PR action and were unlikely to be functionally redundant.

10.3 The Human Progesterone A and B Forms Manifest Distinct Activities Within Target Cells

Although the existence of two forms of PR had been reported in several species, it was not until their respective cDNAs were cloned and incorporated into reconstituted transcription systems that the significance of this finding was appreciated. The initial studies examined the transcriptional activity of the chicken PR-A and PR-B and determined that whereas both isoforms of cPR were capable of activating transcription, they each demonstrated a unique promoter specificity (Gronemeyer et al. 1987; Tora et al. 1988; Dobson et al. 1989). Several years later a similar analysis was performed using the cloned hPR-A and hPR-B

Fig. 1a–c. Differential transcriptional activities of human progesterone receptor-B (*hPR-B*), human progesterone receptor-A (*hPR-A*), chicken progesterone receptor-A (*cPR-A*). **a** Sequence similarities among PR isoforms. The DNA sequences of the human and the chicken isoforms of the PR were obtained from GenBank. Regions of amino acid similarities between hPR-A and cPR-A were determined using the DNA Strider and LALNVIEW programs. The regions of least homology are located upstream of the unique PmL1 restriction site present in both receptors (55% similarity, 30% identity). Regions of high homology are found downstream of the PmL1 restriction site (90% similarity, >72% identity). The amino acid sequence of the B isoform of the hPR is also shown. *HBD*, hormone-binding domain; *AF-1–3*, activation function-1–3; *DBD*, DNA-binding domain. **b** HeLa cells or HepG2 cells (**c**) were transiently transfected with increasing concentrations of vectors expressing hPR-B, hPR-A, or cPR-A ranging from 1 to 5 X, where X represents the respective concentration for each receptor corrected for molarity (where X = 0.131 μg for hPR-B, 0.12 μg for hPR-A, and 0.115 μg for cPR-A). The transcriptional activity was measured 24 h after the addition of 10^{-7} M R5020. In these experiments, PR transcriptional activity was assayed on a PRE2-TK-LUC promoter (1.5 μg). Transfections were normalized for efficiency using 0.05 μg of an internal β-galactosidase control plasmid (pBKC-βgal). Luciferase activity (*Luc. activity*) was normalized to β-galactosidase activity. The total concentration of cytomegalovirus (CMV) promoter was kept constant throughout the experiment by including the appropriate amount of a CMV-based control plasmid (pBK-Rev-TUP1). The total amount of DNA per triplicate was 3.0 μg. Each data point represents the average of triplicate determinations of the transcriptional activity under the given experimental conditions. The average coefficient of variation at each hormone concentration was less than 12%. *NR*, no receptor; –, absence of hormone; +, presence of hormone. (Reproduced with permission from Giangrande et al. 1997)

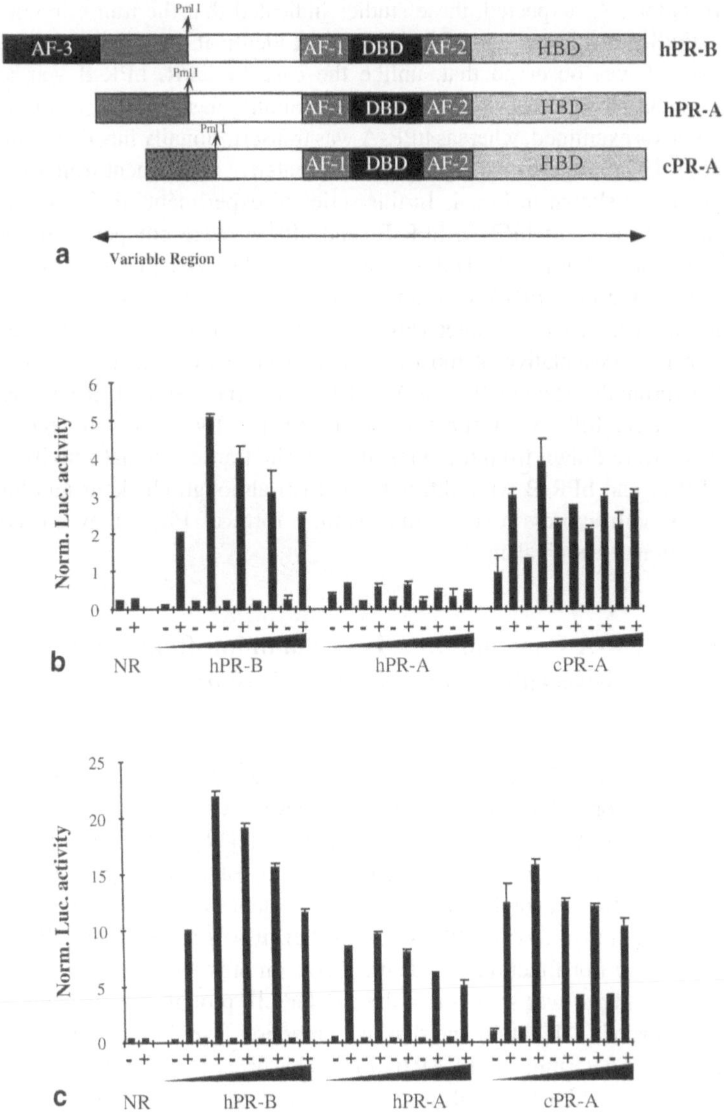

Fig. 1a–c. Legend see p. 182

receptors. As expected, these studies indicated that the transcriptional activities of hPR-A and hPR-B were not identical. Surprisingly, however, it was observed that, unlike the case for cPR, hPR-B was an efficient activator of reporter genes containing classical PREs in most contexts examined, whereas hPR-A was transcriptionally inactive (Tung et al. 1993; Vegeto et al. 1993). A representative experiment from these studies is shown in Fig. 1. In this series of experiments the transcriptional activity of hPR-A, hPR-B, and cPR-A were compared in two different cell lines. In HeLa cells, it was observed that hPR-B and cPR-A, but not hPR-A, can function as efficient activators of a progesterone-responsive promoter (Fig. 1b). The result obtained in this cell line is representative of most cells examined. However, in a few cell backgrounds, HepG2 (Fig. 1c) and Ros17.2 (not shown), it was observed that hPR-A can function as an activator. Two important conclusions were drawn from these results: (1) The transcriptional activity of hPR-A and hPR-B were different, and (2) although chickens and humans both express a structurally similar form of PR-A they are not functionally equivalent.

10.4 hPR-A Functions as a Transdominant Repressor of Steroid-Receptor Transcriptional Activity

There are several examples of single genes encoding multiple forms of a single transcription factor (Desombres and Schibler 1991; Foulkes and Sassone-Corsi 1992). One of the most relevant, from the perspective of PR, was the finding that the transcriptional regulator LAP was coexpressed in target cells with a shorter form, LIP. These proteins are produced from a single mRNA as a consequence of alternate initiation of translation. As with hPR-A and hPR-B, it was shown that LIP and LAP were not functionally identical (Desombres and Schibler 1991). Specifically, it was shown that the shorter LIP protein was actually an inhibitor of LAP action and that its expression level in the cell was an important determinant of LAP biology. This observation prompted us to test whether hPR-A could function similarly and regulate hPR-B transcriptional activity. This analysis revealed that, in cells where hPR-A had no inherent transcriptional activity, it actually functioned as a ligand-dependent inhibitor of hPR-B transcriptional activity (Vegeto et al.

1993). This result implied therefore that the primary role of hPR-A was to regulate hPR-B transcriptional activity and that alterations in the relative expression of the two isoforms would be important in determining PR pharmacology.

Although the role of hPR-A as a modulator of hPR-B signaling was predictable based on the established paradigm of LAP and LIP, it was a surprise when it was observed that hPR-A also functioned to regulate the activity of other steroid receptors. This awareness came from a series of control experiments which were designed to test the specificity of hPR-A's inhibitory activity. However, the surprising result of these studies was that co-expression of hPR-A led to the inhibition of the transcriptional activity of the steroid hormone receptors [estrogen receptor (ER); androgen receptor (AR); glucocorticoid receptor (GR); and mineralocorticoid receptor (MR)] but not of other members of the nuclear receptor superfamily (McDonnell and Goldman 1994; McDonnell et al. 1994; Tzukerman et al. 1994; Wen et al. 1994). Of particular importance was the finding that hPR-A, in the presence of agonists, or antagonists, could completely inhibit estradiol-activated ER-mediated transcriptional activity (McDonnell and Goldman 1994). A representative example of this activity is shown in Fig. 2. In cells expressing ER under the control of a constitutive viral promoter, we observed that progesterone alone had no effect on estradiol-activated transcription. However, when hPR-A was co-expressed in the cell, it was observed that ER transcriptional activity was suppressed in a ligand-dependent manner. Interestingly, the inhibitory activity of hPR-A was induced by either agonists or antagonists (Vegeto et al. 1993; McDonnell and Goldman 1994; Wen et al. 1994). The specificity of this response was demonstrated by showing that hPR-B was unable to modulate ER transcriptional activity (McDonnell and Goldman 1994). Although these results were initially puzzling they suggested a mechanism by which the PR and ER signaling pathways could be integrated. Physiologically, progesterone is the natural antagonist of estrogen-mediated processes, an activity which most believe to be related to its ability to down-regulate ER. However, in these studies where ER is expressed from a constitutive viral promoter, we can demonstrate an antiestrogenic activity of PR ligands under conditions where ER expression is not effected. Consequently, these findings indicate that, in addition to receptor regulation, a more complicated molecular mechanism permitting PR–ER crosstalk is

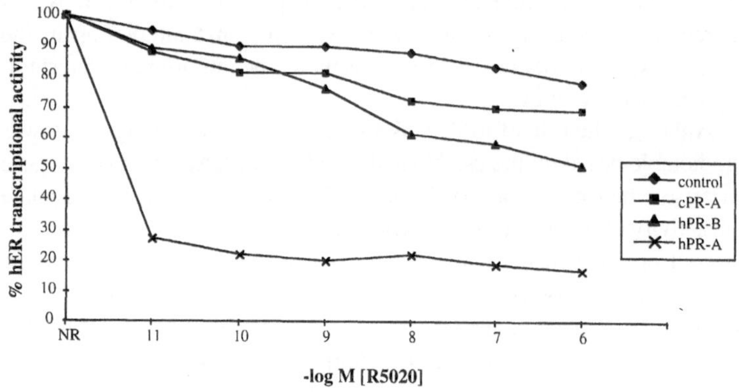

Fig. 2. Transdominant repressor effect of human progesterone receptor A (*hPR-A*), but not human progesterone receptor B (*hPR-B*), or chicken progesterone receptor A (*cPR-A*), on human estrogen receptor (*hER*) transcriptional activity. HeLa cells were transiently transfected with vectors expressing hER alone or in combination with a vector expressing hPR-A (pBK-hPR-A), hPR-B (pBK-hPR-B), or cPR-A (pBK-cPR-A), respectively. The vector pBK-hPR-B was modified by mutating the second in-frame ATG which could potentially yield the A-form of PR. This allows for the expression of hPR-B alone. The transcriptional activity of these constructs was measured following the addition of 10^{-7} M 17β-estradiol alone or in combination with increasing concentrations of R5020 (ranging from 10^{-11} to 10^{-6} M), a progesterone synthetic analogue. In these experiments, ER transcriptional activity was assayed on an ERE_3-TATA-LUC reporter. Transfections were normalized for efficiency using an internal β-galactosidase control plasmid (pBK-βgal). The data are presented as percentages of activation, where 100% represents a measure of 17β-estradiol-dependent transactivation by hER in the presence of a control vector, pBK-Rev-TUP1 (*diamonds*), or in the presence of hPR-A (*X*), hPR-B (*triangles*), or cPR-A (*squares*), respectively, all in the absence of added PR ligands. This value is independently calculated for each data point. Each data point represents the average of triplicate determinations of the transcriptional activity under given experimental conditions. The average coefficient of variation at each hormone concentration was <10%. (Reproduced with permission from Giangrande et al. 1997)

also likely to be operative. This hypothesis, based primarily on the *in vitro* data presented above, was supported by studies in the rat performed by other groups where it was shown that the PR antagonist, RU486, could inhibit estradiol-activated transcription without altering ER expression levels (Kraus and Katzenellenbogen 1993). It appeared from these data, therefore, that the A form of PR was a key modulator of ER pharmacology. Although it was shown that GR, MR, and AR transcriptional activity was also affected by hPR-A expression, the physiological significance of this activity awaits a determination of whether these receptors, and hPR-A, are co-expressed in target cells. Based on our current knowledge of tissue distribution of the steroid hormone receptors, it is likely that the most significant activity of hPR-A will be observed in reproductive tissues where ER- and hPR-B-mediated signaling pathways are both operative.

10.5 The Mechanism of hPR-A-Mediated Inhibition of Steroid Receptor Signaling

The observation that hPR-A inhibited the transcriptional activity of all members of the steroid hormone receptor superfamily suggested that hPR-A was able to inhibit some common pathway or process. Based on the state of the art at the time we considered that "squelching" or titration of a limiting factor required for steroid receptor action was the most likely mechanism. This possibility was addressed in a series of studies which probed the mechanism of hPR-A-mediated inhibition of ER transcriptional activity (Wen et al. 1994). In the absence of expressed hPR-A the transcriptional activity of ER was not affected by the addition of any PR agonists or antagonists (Fig. 3). In contrast, however, the introduction of hPR-A permitted the antiestrogenic activity of these compounds to be manifested. In support of a "squelching" model it was shown that the ability of hPR-A to inhibit hER transcriptional activity was related in a direct manner to its expression level (Fig. 3). However, when we tried to reverse the inhibitory activity of hPR-A by overexpressing hER it was determined that inhibition appeared to depend on the absolute expression level of hPR-A within the cell (Fig. 4). This indicated that squelching, a competitive phenomenon, was not involved, but rather that hPR-A was functioning in an indirect, non-competitive

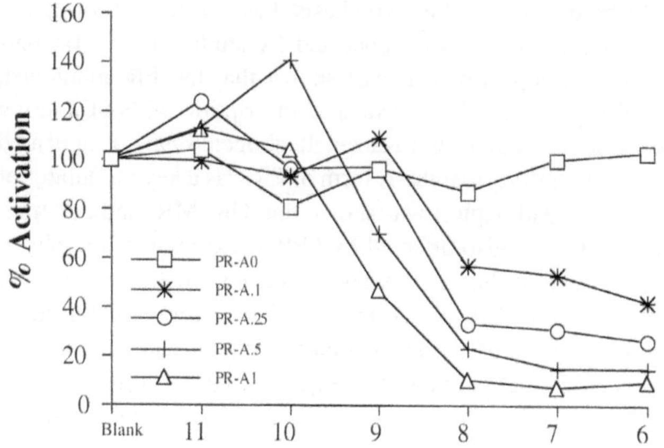

a Progesterone -(logM)

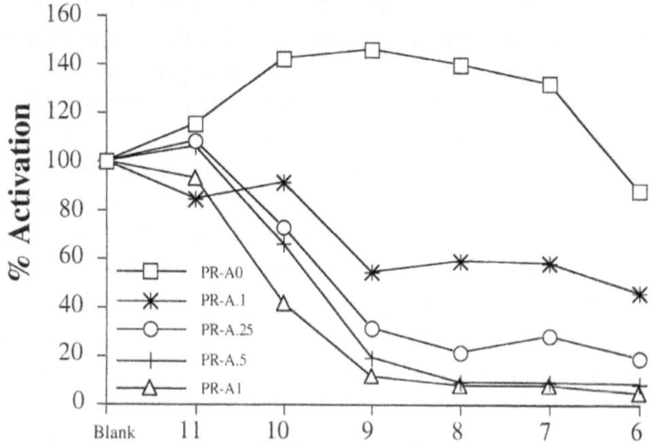

b RU486 -(logM)

Fig. 3a,b. Legend see. p. 189

manner (Wen et al. 1994). In support of this mechanistic distinction, it has been observed that receptor squelching can be overcome, at least in part, by overexpression of either the steroid receptor co-activator, SRC-1, or the receptor potentiating factor, RPF-1 (Oñate et al. 1995; Imhof and McDonnell 1996). Importantly, however, this manipulation has no effect on the ability of hPR-A to function as a transdominant inhibitor (our unpublished results). We have incorporated data into a working model to explain transdominant inhibitory activity of hPR-A on hER transcriptional activity (Fig. 5). Specifically, we propose that in order for hER to activate transcription it must contact the general transcription apparatus (GTA). In this model, it is implied that this interaction requires the recruitment of at least one, but possibly several, co-activators (X and Y). Data from our studies and those of others suggest that hPR-A can prevent the assembly of a productive ER-GTA complex in either of two ways. One possibility is that the hPR-A dimer binds to the co-activator Y and prevents the ER–co-activator complex from contacting the transcription apparatus (Fig. 5a). Thus, hER and hPR-A manifest their respective activities through different proteins. A second alliterative suggests that both hER and hPR-A can interact with the same co-activator (X) however, their binding sites are not mutually exclusive. The binding of hPR-A in this model would also prevent the association of

◀ **Fig. 3a,b.** Inhibition of human estrogen receptor (hER) transcriptional activity by progesterone receptor (*PR*) ligands is influenced by hPR-A expression level. The effects of increasing cellular concentrations of hPR-A on hER-mediated transcriptional activity were measured in CV-1 cells. An expression vector encoding hER (pRST7hER) (5 µg/ml) was transfected into CV-1 cells alone or in the presence of different concentrations of an hPR-A expression plasmid as indicated. All transfection mixes contained an ERE-TK-LUC reporter (10 µg/ml) (25) and pCH110 (5 µg/ml) as an internal control. The transcriptional activity under these conditions was measured following the addition of 10^{-7} M 17-β-estradiol alone or estradiol in the presence of increasing concentrations of progesterone (**a**), or RU486 (**b**), as indicated. Following incubation, cells were harvested and luciferase and β-galactosidase activities were measured. The data are presented as percentages of activation, 100% representing the activity of hER in each condition in the absence of any added PR ligand. Each datum point represents the average of triplicate determinations of the transcriptional activity under a given experimental condition. The average coefficient of variation at each hormone concentration was <15% in this experiment. (Redrawn with permission from Wen et al. 1994)

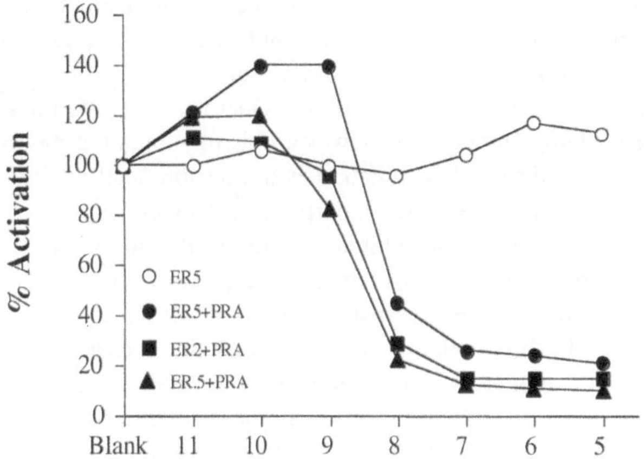

a Progesterone (-logM)

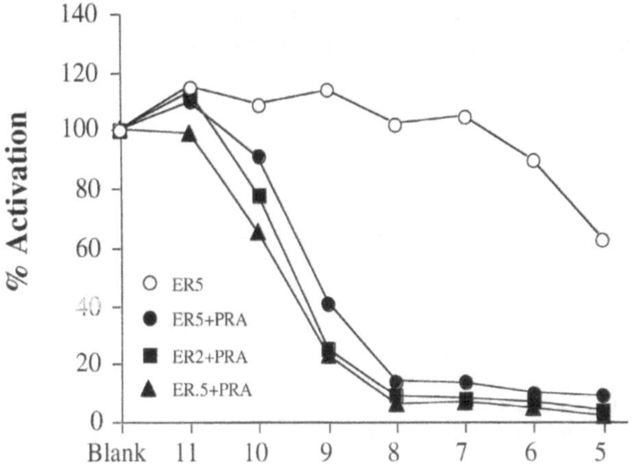

b RU486(-logM)

Fig. 4a,b. Legend see p. 191

the ER–co-activator complex with the general transcription apparatus (Fig. 5b). In these proposed pathways repression could result from a simple steric inhibition of transcription, the recruitment by hPR-A of a transcriptional repressor, or a combination of both. The identification of the proteins which specifically interact with hER and hPR in these complexes will be required to resolve this issue.

◀ **Fig. 4a,b.** Inhibition of human estrogen receptor (*hER*) transcriptional activity by subsaturating concentrations of human progesterone receptor A (*hPR-A*) is independent of hER expression level. Monkey kidney CV-1 cells were transiently transfected with increasing concentrations of an hER expression plasmid (as indicated) alone or in the presence of a vector expressing hPR-A. The concentration of hPR-A expression vector (0.5 μg/ml) was shown previously to be submaximal for hPR-A-mediated repression of hER activity. Each transfection condition included a mouse mammary tumor virus-estrogen response element-luciferase (MMTV-ERE-LUC) reporter plasmid (10 μg/ml) and plasmid pCH110 (5 μg/ml) as an internal control. The transcriptional activity in these setups was measured following the addition of 10^{-7} M 17-β-estradiol alone or estradiol in the presence of increasing concentrations of progesterone (**a**), or RU486 (**b**), as indicated. Following incubation, the cells were harvested and luciferase and β-galactosidase activities were measured. The data are presented as percentages of activation; the 100% value represents the activity of hER in each condition in the absence of any added PR ligand. To confirm that the range of ER expression vector chosen allowed an examination of hPR-A activity in the presence of subsaturating and saturating levels of ER, we calculated the fold induction by estradiol in each transfection. The fold inductions in the experiments detailed in (**a**) and (**b**) were as follows: 5 μg of ER expression vector, 31, 26, and 35, respectively; 2 μg of hER expression vector, 28, 24, and 33, respectively; 0.5 μg of hER expression vector, 18, 16, and 24, respectively. Each datum point shown represents the average of triplicate determinations of the transcriptional activity under a given experimental condition. The average coefficient of variation at each hormone concentration was <15% in this experiment. The data presented represent several individual experiments

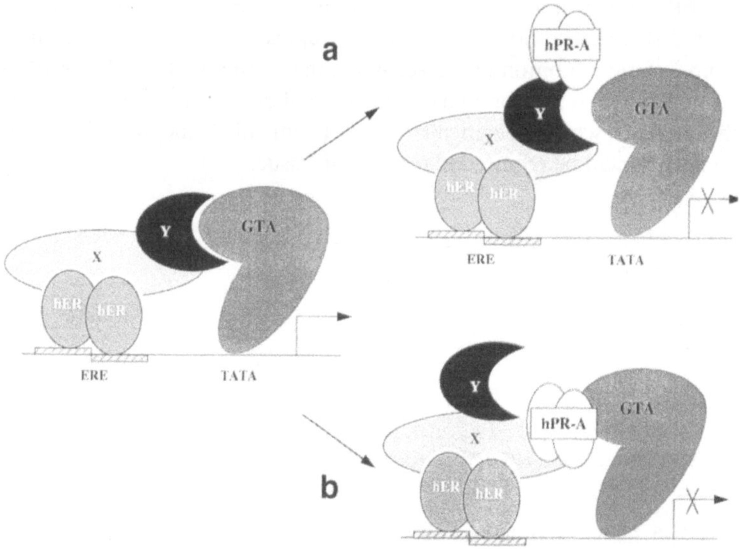

Fig. 5a,b. Human progesterone receptor A (*hPR-A*) functions as a transdominant inhibitor of human estrogen receptor (*hER*) function. Two working models detailing possible mechanisms by which hPR-A can function as a transdominant repressor of hER function are presented. Our data and that of others suggest that ER enhances gene transcription by recruiting one or more transcriptional coactivators to a target gene promoter and permitting the indirect interaction of the receptor with the general transcription apparatus (*GTA*). Two similar, though distinct, mechanisms by which hPR-A can inhibit hER transcriptional activity are proposed: **a** One possibility is that hPR-A and hER do not interact with the same target proteins. Specifically hER, or other steroid receptors, can interact with their respective coactivator proteins (*X*) but this interaction is not productive because it is blocked hPR-A acting through a distinct target protein (*Y*). **b** A second plausible model is that hER and hPR-A interact with distinct sites on the same target protein, but that binding of the two proteins is not mutually exclusive. However, the interaction of hPR-A with this target inhibits the positive effect of hER by interrupting the interaction of the receptor–coactivator complex with the transcription apparatus. *ERE*, estrogen response element

10.6 Identification of a Distinct Inhibitory Domain Within hPR-A Required for its Ability to Inhibit Steroid Receptor-Mediated Transcription

It has been assumed that the 164 amino acid hPR-B specific region within PR contains a domain responsible for the functional differences between the two PR isoforms. Indeed, some work has suggested that a distinct activation domain, B upstream sequence (BUS), does in fact exist within this domain (Sartorius et al. 1994). Unfortunately, however, the role of BUS is unclear as it has only been shown to function as an autonomous activator when fused to the PR DNA-binding domain and not other heterologous DNA-binding domains. It has yet to be determined if it will actually function as an activator within the context of the full-length PR-B. Another way of explaining the differential activities of hPR-A and B is that both receptors have all the information required to permit the receptors to function as transcriptional activators but that in the context of hPR-A this function is suppressed. This hypothesis implies that sequences which are specific to hPR-B may not necessarily constitute a bona fide activation function, but rather they permit the receptor to overcome the activity of an inhibitory domain within the regions of the receptor common to both isoforms (Kastner et al. 1990). To examine this possibility, we took advantage of the observation that cPR-A, although similar in sequence and structure to hPR-A, functioned as an activator and not an inhibitor of progesterone-responsive genes (Krust et al. 1986; Conneely et al. 1987; Misrahi et al. 1987; Giangrande et al. 1997). Using this information, a series of chimeric proteins in which the least conserved regions of the chicken and the human receptors were produced and analyzed in transfected mammalian cells and examined for their ability to inhibit ER-mediated transcription (Fig. 6). This exercise permitted the mapping of a specific inhibitory domain within hPR-A. Specifically, it was demonstrated that the amino terminal 140 amino acids of hPR-A were necessary for its ability to function as a transcriptional inhibitor. Deletion of this region from hPR-A resulted in a receptor mutant which was functionally indistinguishable from hPR-B (Giangrande et al. 1997). In addition, when the corresponding domains within the human and chicken receptors were exchanged it was determined that the chicken–human receptor chimera could activate transcription, whereas the corollary human–chicken receptor chimera was

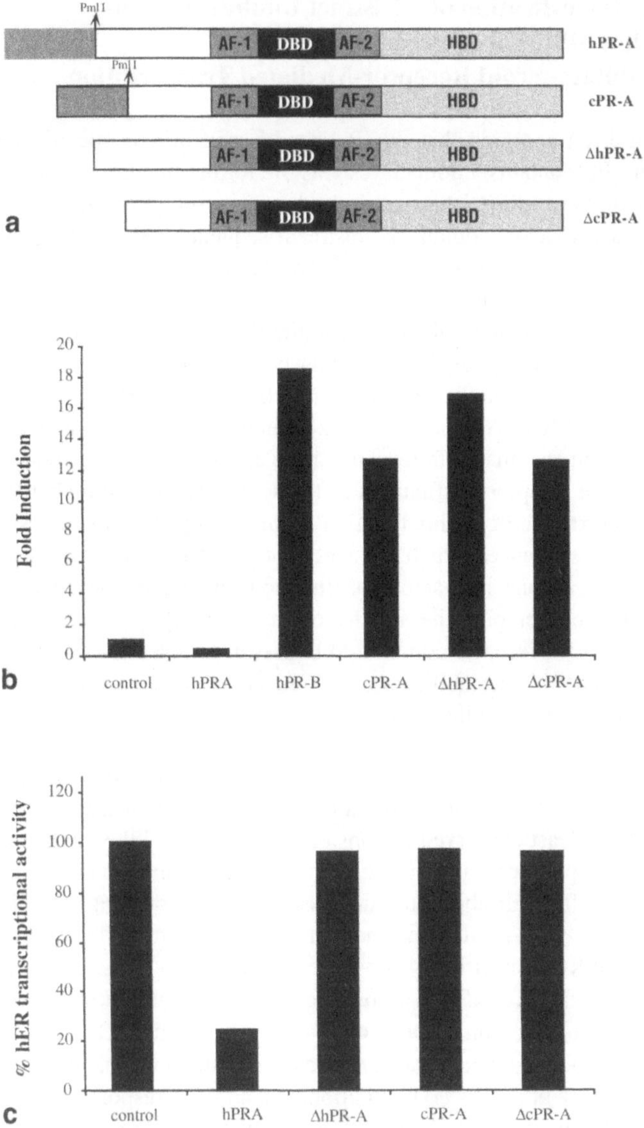

Fig. 6a–c. Legend see p. 195

now a transdominant repressor (Fig. 7). Although these 140 amino acids function as a transferable repression domain they are not able to function as an autonomous repressor when linked to the heterologous GAL4 DNA-binding domain (DBD). Cumulatively, however, these results support the hypothesis that all the sequences required for PR transcriptional activity lie within the region common to hPR-A and hPR-B. However, the inhibitory activity of the amino terminal 140 amino acids of hPR-A masks this activity. We propose, therefore, that the role of the B-specific 164 amino acids is to override the activity of the hPR-A inhibitory domain and permit transcriptional activation.

Fig. 6a–c. Δ human progesterone receptor A (*ΔhPR-A*) is unable to repress human estrogen receptor (*hER*) transcriptional activity. **a** The DNA sequences of hPR-A and chicken PR-A (*cPR-A*) were obtained from GenBank. These represent the full-length sequences of the two receptors. ΔhPR-A and ΔcPR-A, subcloned into pBK-CMV mammalian expression vector, were generated by deleting the 140 amino acids of hPR-A upstream of the PmL1 restriction site and the corresponding 90 amino acids of cPR-A, respectively. *AF-1*/*AF-2*, activation functions 1 and 2; *DBD*, DNA-binding domain; *HBD*, hormone-binding domain. **b** HeLa cells were transiently transfected with vectors expressing hPR-A, ΔhPR-A, cPR-A, or ΔcPR-A, respectively. The transcriptional activity was measured following the addition of 10^{-7} M R5020. A control vector (pBK-RevTUP1) was used to assess the basal level of transcription of the PRE2-TK-LUC reporter. Transfections were normalized for efficiency using the internal pBK-βgal control plasmid. The data are represented as fold induction, a measure of ligand-induced activity divided by basal (no hormone) activity, for each data point. **c** HeLa cells were transiently transfected with vectors expressing hER alone or in combination with a vector expressing hPR-A, ΔhPR-A, cPR-A, or ΔcPR-A, respectively. The transcriptional activity was measured following the addition of 10^{-7} M 17β-estradiol and 10^{-7} M R5020 alone or in combination. In these experiments ER transcriptional activity was assayed on an ERE_3-TATA-LUC promoter. Transfections were normalized for efficiency using an internal β-galactosidase control plasmid. The data are presented as percentages of activation, where 100% represents a measure of 17β-estradiol-dependent transactivation by hER in the presence of control vector alone or in the presence of hPR-A, ΔhPR-A, cPR-A, and ΔcPR-A, respectively, but in the absence of R5020. This value is independently calculated for each data point. Each data point represents the average of triplicate determinations of the transcriptional activity under given experimental conditions. The average coefficient of variation was <10%. (Reproduced with permission from Giangrande et al. 1997)

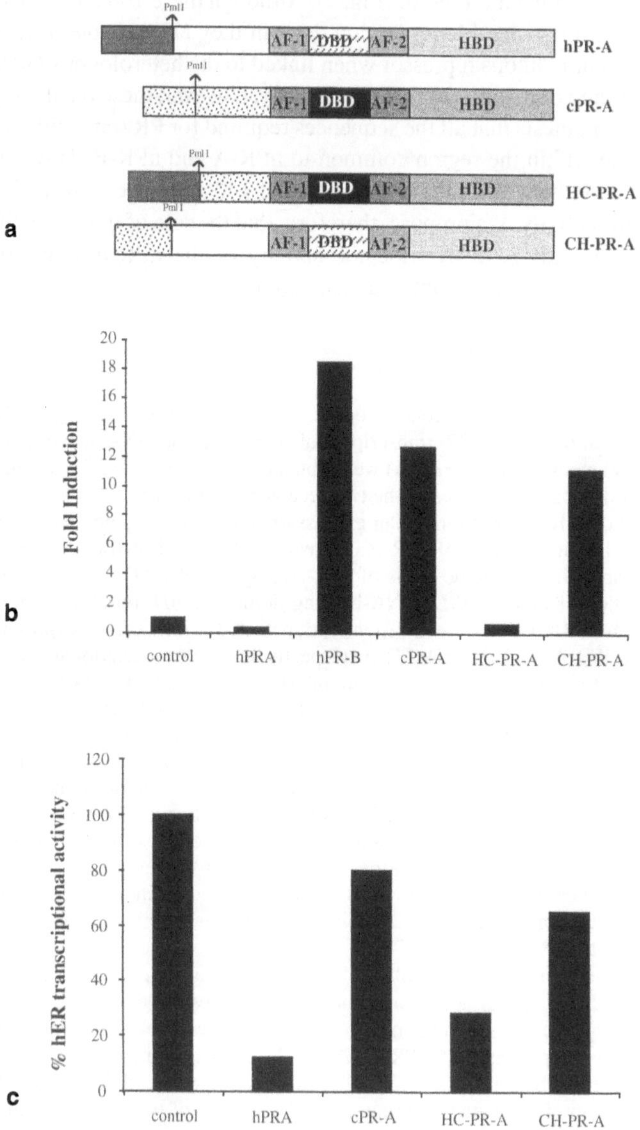

Fig. 7a–c. Legend see p. 197

10.7 Final Comments

The finding that the transcriptional activity of PR-A and PR-B are different and that the influence of PR-A extends to other steroid receptors, begs a redefinition of the term progesterone-responsive genes. Additionally, it suggests that in addition to the expression level of PR-A and PR-B within target cells, alterations in the factors which influence the activities of the two receptor isoforms will have a profound effect on PR pharmacology. These studies have led to the mapping of an inhibitory domain which appears to be responsible for the unique activity of hPR-A. The next step in defining the physiological relevance of this unique activity is to identify the cellular components responsible for distinguishing between hormone-activated hPR-A and hPR-B – an investigative avenue already being carefully explored in several laboratories.

◀ **Fig. 7a–c.** The human chimeric progesterone receptor A construct (*HC-PR-A*) is a potent repressor of human estrogen receptor (*hER*) transcriptional activity in HeLa cells. **a** The human/chicken chimeric constructs (*HC-PR-A* and *CH-PR-A*, respectively) were generated by swapping the N-terminal regions (upstream of the unique PmL1 restriction site, present in both receptors) of the human and chicken receptors. **b** HeLa cells were transiently transfected with vectors expressing human/chicken progesterone receptor A (*hPR-A* and *cPR-A*, respectively) HC-PR-A, and CH-PR-A, respectively. The transcriptional activity of these chimeric constructs was assayed on the PRE_3-TK, progesterone-responsive promoter. The activity was measured after 24-h induction with 10^{-7} M R5020. Fold induction represents the normalized luciferase activity divided by basal (no hormone) activity, for each receptor type after induction with ligand. **c** HeLa cells were transiently transfected with vectors expressing hER alone or in combination with a vector expressing hPR-A, cPR-A, HC-PR-A, and CH-PR-A, respectively. The transcriptional activity was measured following the addition of 10^{-7} M 17β-estradiol and 10^{-7} M R5020 alone or in combination. A control was carried out in the absence of ligands. In these experiments ER transcriptional activity was assayed on an ERE_3-TATA-LUC promoter. Transfections were normalized for efficiency using an internal β-galactosidase control plasmid. The data are presented as percentages of activation, where 100% represents a measure of 17β-estradiol-dependent transactivation by hER in the presence of hPR-A, cPR-A, HC-PR-A, and CH-PR-A, respectively, but in the absence of R5020. This value is independently calculated for each data point. Each data point represents the average of triplicate determinations of the transcriptional activity under given experimental conditions. The average coefficient of variation was < 11% for both experiments. (Reproduced with permission from Giangrande et al. 1997)

Acknowledgements. We thank John Norris for his helpful comments and suggestions and Trena Martelon for editorial assistance. This work was supported by an NIH grant DK50494 (DPM).

References

Brandon DB, Bethea CL, Strawn EY et al (1993) Progesterone receptor messenger ribonucleic acid and protein are overexpressed in human uterine leiomyomas. Am J Obstet Gynecol 169:78–85

Carroll RS, Glowacka D, Dashner K et al (1993) Progesterone receptors in meningiomas. Cancer Res 53:1312–1316

Cavaillès V, Dauvois S, L'Horset F et al (1995) Nuclear factor RIP140 modulates transcriptional activation by the estrogen receptor. EMBO J 14:3741–3751

Christensen K, Estes PA, Oñate SA et al (1991) Characterization and functional properties of the A and B forms of human progesterone receptors synthesized in a baculovirus system. Mol Endocrinol 5:1755–1770

Clark JH, Peck EJ (1979) Female sex steroids: receptors and function. Springer, Berlin Heidelberg New York

Clarke CL, Sutherland RL (1990) Progestin regulation of cellular proliferation. Endocr Rev 11:266–301

Colletta AA, Wakefield LM, Howell FV et al (1991) The growth inhibition of human breast cancer cells by a novel synthetic progestin involves the induction of transforming growth factor beta. J Clin Invest 87:277–283

Conneely OM, Dobson ADW, Tsai M-J et al (1987) Sequence and expression of a functional chicken progesterone receptor. Mol Endocrinol 1:517–525

Conneely OM, Maxwell BL, Toft DO et al (1987) The A and B forms of the chicken progesterone receptor arise by alternate initiation of translation of a unique mRNA. Biochem Biophys Res Commun 149:493–501

DeMarzo AM, Beck CA, Oñate SA et al (1991) Dimerization of mammalian progesterone receptors occurs in the absence of DNA and is related to the release of the 90-kDa heat shock protein. Proc Natl Acad Sci USA 88:72–76

DeMarzo AO, Oñate SA, Nordeen SK et al (1992) Effects of the steroid antagonist RU486 on dimerization of the human progesterone receptor. Biochemistry 31:10491–10501

Desombres P, Schibler U (1991) A liver enriched transcriptional activator protein, LAP and a transcriptional inhibitory protein, LIP are translated from the same gene. Cell 67:569–579

Dobson ADW, Conneely OM, Beattie W et al (1989) Mutational analysis of the chicken progesterone receptor. J Biol Chem 264:4207–4211

Foulkes NS, Sassone-Corsi P (1992) More is better: activators and repressors from the same gene. Cell 68:411–414

Giangrande PH, Pollio G, McDonnell DP et al (1997) Mapping and characterization of the functional domains responsible for the differential activity of the A and B isoforms of the human progesterone receptor. J Biol Chem 272:32889–32900

Gronemeyer H, Turcotte B, Quirin-Stricker C et al (1987) The chicken progesterone receptor: sequence, expression and functional analysis. EMBO J 6:3985–3994

Gronemeyer H, Meyer M-E, Bocquel M-T et al (1991) Progestin receptors: isoforms and antihormone action. J Steroid Biochem Mol Biol 40:271–278

Halachmi S, Marden E, Martin G et al (1994) Estrogen receptor-associated proteins: possible mediators of hormone-induced transcription. Science 264:1455–1458

Hanstein BH, Eckner R, DiRenzo J et al (1996) p300 is a component of an estrogen receptor coactivator complex. Proc Natl Acad Sci USA 93:11540–11545

Hong H, Kohli K, Trivedi A et al (1996) GRIP1, a novel mouse protein that serves as a transcriptional coactivator in yeast for the hormone binding domains of steroid receptors. Proc Natl Acad Sci USA 93:4948–4952

Horwitz KB (1992) The molecular biology of RU486: is there a role for antiprogestins in the treatment of breast cancer? Endocr Rev 13:146–163

Imhof MO, McDonnell DP (1996) Yeast RSP5 and its human homolog hRPF1 potentiate hormone-dependent activation of transcription by human progesterone and glucocorticoid receptors. Mol Cell Biol 16:2594–2605

Jacq X, Brou C, Lutz Y et al (1994) Human $TAF_{II}30$ is present in a distinct TFIID complex and is required for transcriptional activation by the estrogen receptor. Cell 79:107–117

Kastner P, Krust A, Turcotte B et al (1990) Two distinct estrogen-regulated promoters generate transcripts encoding the two functionally different human progesterone receptor forms A and B. EMBO J 9:1603–1614

Kettel LM, Murphy AA, Mortola JF et al (1991) Endocrine responses to long-term administration of the anti-progesterone RU486 in patients with pelvic endometriosis. Fertil Steril 56:402–407

Kraus WL, Katzenellenbogen B (1993) Regulation of progesterone receptor gene expression and growth in the rat uterus: modulation of estrogen actions by progesterone and sex steroid hormone antagonists. Endocrinology 132:2371–2379

Krust A, Green S, Argos P et al (1986) The chicken oestrogen receptor sequence: homology with v-erbA and the human oestrogen and glucocorticoid receptors. EMBO J 5:891–897

Le Douarin B, Zechel C, Garnier J-M et al (1995) The N-terminal part of TIF1, a putative mediator of the ligand-dependent activation function (AF-2) of nuclear receptors, is fused to B-raf in the oncogenic protein T18. EMBO J 14:2020–2033

Lessey BA, Alexander PS, Horwitz KB et al (1983) The subunit characterization of human breast cancer progesterone receptors: characterization by chromatography and photoaffinity labelling. Endocrinology 112:1267–1274

Lundgren S (1992) Progestins in breast cancer treatment. Acta Oncol 31:709–722

McDonnell DP (1995) Unraveling the human progesterone receptor signal transduction pathway: insights into antiprogestin action. Trends Endocrinol Metab 6:133–138

McDonnell DP, Goldman ME (1994) RU486 exerts antiestrogenic activities through a novel progesterone receptor A form-mediated mechanism. J Biol Chem 269:11945–11949

McDonnell DP, Shahbaz MS, Vegato E et al (1994) The human progesterone receptor A-form functions as a transcriptional modulator of mineralocorticoid receptor transcriptional activity. J Steroid Biochem Mol Biol 48:425–432

Mengus G, May M, Jacq X et al (1995) Cloning and characterization of $hTAF_{II}18$, $hTAF_{II}20$ and $hTAF_{II}28$: three subunits of the human transcription factor TFIID. EMBO J 14:1520–1531

Misrahi H, Atger M, d'Auriol L et al (1987) Complete amino acid sequence of the human progesterone receptor deduced from cloned cDNA. Biochim Biophys Res Commun 143:740–748

Oñate SA, Tsai S, Tsai M-J et al (1995) Sequence and characterization of a coactivator for the steroid hormone receptor superfamily. Science 270:1354–1357

Poisson M, Pertuiset BF, Hauw JJ et al (1983) Steroid hormone receptors in human meningiomas, gliomas and brain metastases. J Neuro Oncol 1:179–189

Sartorius CA, Melville MY, Hovland AR et al (1994) A third transactivation function (AF3) of human progesterone receptors located in the unique N-terminal segment of the B-isoform. Mol Endocrinol 8:1347–1360

Schwerk C, Klotzbucker M, Sacks M et al (1995) Identification of a transactivation function in the progesterone receptor that interacts with the TAF11 110 subunit of the TFIID complex. J Biol Chem 270:21331–21338

Tora L, Gronemeyer H, Turcotte B et al (1988) The N-terminal region of the chicken progesterone receptor specifies target gene activation. Nature 333:185–188

Tung L, Mohamed MK, Hoeffler JP et al (1993) Antagonist-occupied human progesterone B-receptors activate transcription without binding to progesterone response elements and are dominantly inhibited by A-receptors. Mol Endocrinol 7:1256–1265

Tzukerman MT, Esty A, Santiso-Mere D et al (1994) Human estrogen receptor transcriptional capacity is determined by both cellular and promoter context and mediated by two functionally distinct intramolecular regions. Mol Endocrinol 8:21–30

Vegeto E, Allan GF, Schrader WT et al (1992) The mechanism of RU486 antagonism is dependent on the conformation of the carboxy-terminal tail of the human progesterone receptor. Cell 69:703–713

Vegeto E, Shahbaz MM, Wen DX et al (1993) Human progesterone receptor A form is a cell- and promoter-specific repressor of human progesterone receptor B function. Mol Endocrinol 7:1244–1255

Voegel JJ, Heine MJS, Zechel C et al (1996) TIF2, a 160 kDa transcriptional mediator for the ligand-dependent activation function AF-2 of nuclear receptors. EMBO J 15:3667–3675

Wen DX, Xu YF, Mais DE et al (1994) The A and B isoforms of the human progesterone receptor operate through distinct signaling pathways within target cells. Mol Cell Biol 14:8356–8364

11 Oestrogen Receptor Expression Mutants and Variants in Tamoxifen-Resistant Breast Cancer

C.M.W. Chan and M. Dowsett

11.1 Introduction

The large majority of breast cancer patients will receive tamoxifcn at some time during their treatment. It is now known that tamoxifen can enhance survival in patients with early breast cancer but many of these will relapse at a later stage with disease which has become resistant to this compound. Patients who receive tamoxifen with metastatic disease have overall a 30% chance of showing an objective response. Approximately 20% show stabilised disease whilst the remainder progress, exhibiting what is called intrinsic or de novo resistance. The majority of these patients are oestrogen receptor (ER)-negative and few will respond to treatment with a second line endocrine agent (Saez and Osbourne 1989). Almost all of the initially responding patients will eventually relapse showing what is called acquired resistance. Given the mechanism of action of tamoxifen as a competitive oestrogen antagonist, it is not surprising that virtually all of the responding patients show ER-positive disease. An understanding of the mechanism of tamoxifen resistance will help to advance our development of better agents for breast cancer treatment. Several mechanisms have been hypothesised as to how resistance could arise. These include loss or mutation of ER; post-receptor alterations, such as changes in phosphorylation pathways; changes in growth factor production/sensitivity or paracrine cell–cell interactions; altered uptake, retention or metabolism of the antioestrogen (Katzenebollenbogen et al. 1997). The majority of studies have focused on the nature of ER expression in tamoxifen resistance. This article seeks to summarise studies of expression and function of ER and its role in generating tamoxifen-resistant human breast cancer. There are two forms of the ER, ERα and ERβ. All of the studies described in this chapter were performed with the ERα receptor.

11.2 Oestrogen Receptor Expression

During the early 1980s, a series of studies of tamoxifen-treated tumours reported that resistance was associated with decreased ER levels using ligand-binding assays (Allegra et al. 1980; Nomura et al. 1985). However, these findings are mainly in conflict with the results of recent studies employing monoclonal antibodies to the receptor (McCarty et al. 1986). It seems likely that these earlier observations were aberrant as a result of tamoxifen interfering in the ligand-binding assays.

We have previously reported our findings in which oestrogen expression was measured by immunocytochemical techniques (Johnston et al. 1995). Pre-treatment biopsies were compared with those obtained at progression of the tumour whilst patients remained on tamoxifen treatment. In 20 pairs of samples from patients with de novo resistance, only 3 tumours (15%) were ER-positive at pre-treatment. There was no significant change in ER, progesterone receptor (PR) or pS2 levels in this group. In 18 pairs of samples from patients with acquired tamoxifen resistance, 16 (89%) showed ER positivity post-treatment. Whilst there appeared to be a trend downward in ER expression, this was not statistically significant and it was clear that the large majority of patients retained ER levels consistent with hormone-responsive breast cancer despite the tumour not being tamoxifen resistant. Similarly there was no change in PR or pS2 levels providing some evidence of the functionality of the persistent ER.

These data were in contrast to those from 34 matched samples of patients treated with tamoxifen in the adjuvant context (i.e., in patients receiving tamoxifen after surgical excision of stage I or II breast cancer). In this circumstance there were significant changes in ER and PR levels: 53% of patients were ER-positive pre-treatment compared to 29% post-treatment and the mean ER score fell from 103 to 57 ($p = 0.0002$). Similarly, only 12% of patients were PR-positive compared with 38% pre-treatment. Interestingly, there was no significant change in pS2 levels. In interpreting these data, it is important to be aware that the pre-treatment sample was compared with a recurrent lesion from a different site taken some years after removal of the initial region. Thus there is a potential for changes in status to occur as a result of either phenotypic drift with time or selection of a more aggressive phenotype through the metastatic process. This type of change, in the absence of intervening therapy, has been noted to occur by Kuukasjarvi et al. (1996) who found that only 23 of 35 ER-positive tumours at presentation retained ER at recurrence. These data are in conflict to some extent with the overall view that is held by some (Robertson 1996) that ER is a stable phenotype. Whilst the results from Kuukasjarvi et al. (1996) may be extreme they do emphasised that such change in status can occur independent of hormonal treatment.

We would conclude that: (a) Lack of ER is a major cause of de novo resistance, (b) the loss of ER is not a major cause of acquired resistance, and (c) loss of ER in adjuvant resistance is, at least partly, due to tamoxifen-independent phenotypic drift of metastatic selection.

11.3 Oestrogen Receptor Function

We proceeded to ask the question of whether the ER which persists in tamoxifen-resistant disease is functionally intact, and have taken the first step in this by looking at its ability to bind to oestrogen response elements (EREs) in the form of a gel shift assay. It was found that 37 out of 66 tumours with moderate ER concentrations (10–99 fmol/mg protein) showed positive binding as compared with 37 out of 45 tumours with greater levels of ER. A total of 12 out of 16 (67%) ER-positive tumours which were either PR- or pS2-positive showed such binding, as did nine out of ten ER-positive tumours which were negative for both PR and pS2 (Johnston et al. 1997). Thus binding to DNA was observed even in those tumours in which this protein expression did not confirm the intact nature of the ER mechanism.

11.4 Oestrogen Receptor Mutants

In ER knockout mice, the complete absence of a functional ER leads to adult female mice with only vestigial ducts present at the nipples (Korach 1994), but the impact of specific germ-line ER mutations in transgenics have not been assessed. Although numerous germ-line mutations of androgen receptor have been described with consequent androgen-resistance syndromes (Hughes and O'Malley 1991), an analogous syndrome of oestrogen resistance has been described in only one male (Smith et al. 1994).

Point mutations of mouse ER, engineered between residues 538 and 552, were shown to have reduced transcriptional activation (Mahfoudi et al. 1995). Although hormone- or DNA-binding were not significantly affected, the pharmacological behaviour of oestrogen antagonists, tamoxifen and ICI 164384, was altered dramatically. However, unlike the other mutants in this region, oestradiol was shown to act as an antagonist with a Leu540Gln mutant (Montano et al. 1996). In addition, a naturally occurring Tyr537Asn mutant ER, with oestradiol-independent transcriptional activity and virtually unaffected by tamoxifen or ICI 164384, has been isolated from metastatic breast cancer tumours (Zhang et al. 1997). If this mutant is present in other metastatic breast cancer, it might contribute to tamoxifen resistance. An Asp351Tyr mutant ER was also found in an acquired tamoxifen-resistant breast cancer xenograft (Wolf and Jordan 1994). This mutation is located in the ligand-binding domain and is some distance from the other mutations found in the transcription activation function (AF)-2 region (Mahfoudi et al. 1995; Montano et al. 1996). However, although we and others have found no association between ER mutations and hormone receptor status or tamoxifen resistance (Andersen et al. 1997; Roodi et al. 1995), a mutation at codon 325 of the ER protein has been reported as being associated with a family history of breast cancer (Roodi et al. 1995). Additionally, a "super-active" ER mutation has been described which leads to a 200-fold increase in the proliferative response of MCF-7 cells to oestradiol (Witschke et al. 1996). Despite the apparent frequent occurrence of this in benign proliferative disease, this mutation has not been reported in breast carcinomas.

In conclusion, while some ER mutations might be involved in the aetiology of breast cancer this appears to be rare and such mutations do not appear to be associated with the occurrence or acquisition of hormone resistance.

11.5 Oestrogen Receptor Variants

By contrast, numerous variant mRNA forms have been described and are often found in breast carcinomas. Mostly they may take the form of exon deletions, but there are mRNA variants with duplicated exons (Murphy et al. 1996) or with insertions of ER intron sequence (Wang et al. 1997). Furthermore, we have also described two cryptic splice variants (Daffada et al. 1995; Chan and Dowsett 1997). Hence most variant mRNAs are partly translated out of frame until a nonsense codon terminates translation, leading to a truncated ER protein with some non-ER C-terminal sequence. For instance, translation of the Δ5 mRNA variant results in a 40-kDa truncated protein, lacking the ligand-binding domain (Fuqua et al. 1991). Only two in-frame variants have been described in which exon 3 or 4 are deleted (Δ3 and Δ4) such that the predicted protein products would be deficient only in those regions. Although for the majority of products function has not been described, the Δ5 variant has been found to be dominantly positive in some model systems (Fuqua et al. 1991) and dominant negative activity has been suggested for both Δ3 and Δ7 variants (Wang and Miksicek 1991; Fuqua et al. 1992). As the Δ5 variant has been the most widely studied, it will be used to illustrate the possible importance of mRNA variants in breast cancer resistance. A more complete list of the other mRNA variants found can be obtained elsewhere (Castles and Fuqua 1996; Pfeffer et al. 1995; Murphy et al. 1997).

It was proposed that a constitutively expressed Δ5 variant ER might explain the existence of the uncommon ER–ve/PR+ve breast carcinoma

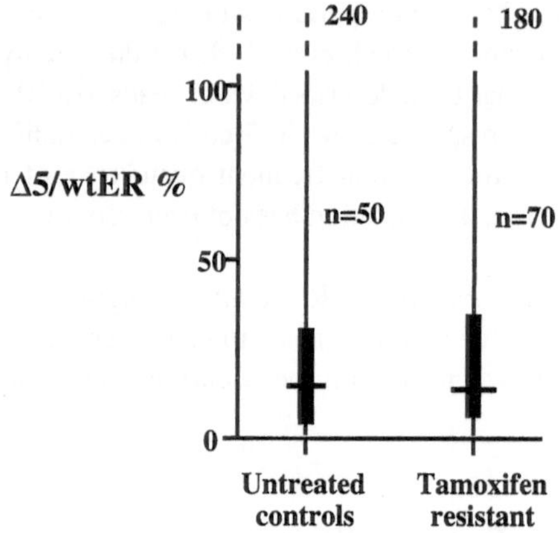

Fig. 1. Δ5 wild-type oestrogen receptor (*Δ5/wtER*) ratios in 50 untreated breast carcinomas compared with 70 tamoxifen-resistant breast carcinomas. *Narrow bars*, range; *solid bars*, interquartile range; *horizontal line*, median

phenotype (Fuqua et al. 1991). Δ5 mRNA levels were higher than wild-type in three of four ER–ve/PR+ve breast carcinomas, whilst they were lower in all five of the ER+ve/PR+ve tumours examined and the Δ5 variant was constitutively activated when expressed in yeast cells (Fuqua et al. 1991; Castles et al. 1993). Furthermore, when transfected into MDA-MB231 breast cancer cells, the variant activated an ERE-tk-luciferase plasmid with about 60% of the efficiency of wild-type receptor (Fuqua et al. 1995) and this activity was independent of oestrogen ligand and unaffected by oestrogen antagonists. MCF-7 cells expressing Δ5 variant mRNA grew faster and expressed more PR than those that did not (Fuqua et al. 1995). In addition, these cells were 4-hydroxytamoxifen-resistant, although the pure antioestrogen ICI 182780 suppressed their growth. The reason for this dichotomy between the effects with the two antioestrogens is unclear, but it is possible that the variant may dimerise with wild-type receptor which is known to be markedly down-regulated by ICI 182780 but not tamoxifen (DeFriend et al. 1994). Importantly, the Δ5 protein is the only variant protein which can be detected in an extract of the ER–ve, weakly PR+ve BT20 breast cancer cell line (Castle'et al. 1993). This remains the only conclusive evidence of the existence of any naturally occurring variant protein.

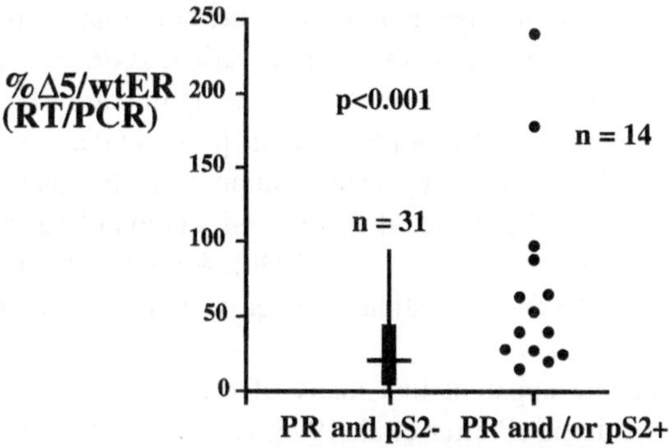

Fig. 2. Δ5 wild-type oestrogen receptor (*Δ5/wtER*) ratio in 31 ER-progesterone receptor (*PR*)-pS2– breast carcinomas compared with 14 ER-PR+ or pS2+ breast carcinomas (individual ratios). *Narrow bars*, range; *solid bars*, interquartile range; *horizontal line*, median; *RT/PCR*, reverse transcriptase polymerase chain reaction

Tamoxifen resistance, however, was found not to be associated with the presence of Δ5 variant mRNA in tumour samples (Daffada et al. 1995), although it was more commonly expressed in ER–ve tumours which expressed PR or pS2 (Figs. 1, 2). In addition, a recent study with Δ5 ER-transfected MCF-7 cells has failed to confirm the earlier results described above (Rea and Parker 1996).

Shorter truncations of the ER receptor, at exon 2 or exon 3, appear not to be functional even if all or part of the DNA-binding sequence is present (Dotzlaw et al. 1992), although the Δ3 variant could repress the oestrogen-induced activity of an ERE-tk chloramphenicol acethyltransferase (CAT) promoter in HeLa cells (Wang and Miksicek 1991). However, others were unable to confirm this (McGuire et al. 1991). Similarly, although the Δ4 ER variant in human embryocarinoma and choriocarcinoma cells contained both of the DNA-binding zinc fingers, unlike the Δ5 variant, there was no transactivating function nor can it interfere with the wild-type receptor activity (Koehorst et al. 1993).

In contrast to the Δ3 variant, the transcriptionally inactive Δ7 variant acted as a dominant negative to down-regulate the transactivational activity of wild-type ER when co-expressed in yeast (Fuqua et al. 1992). About half of the ER+ve/PR–ve tumours expressed this variant, suggesting that this might have an impact on the hormonal responsiveness of these tumours. Similarly, the cryptic variant Δ4/6 also acted as a dominant negative and the expression was significantly higher in tumours which are ER+ve/PR or pS2–ve. The prospect that several variant messages may be expressed by a single tumour in addition to wild-type ER has initiated studies to see if there is association of variant profiles with tumour resistance (Leygue et al. 1996). Although the data indicate that there might be an association, a larger study is required to confirm these findings.

It has become apparent that many of these variants also exist in normal breast tissue (Pfeffer et al. 1995; Leygue et al. 1996). Furthermore, we have also established that Δ5 and the cryptic variant Δ4/6 message is also expressed along with the wild-type receptor in normal liver, endometrium, as well as normal breast and breast cancer (Chan and Dowsett 1997; Daffada and Dowsett 1995). Expression also appears to be tissue-specific since the exon 4/7 cryptic splice variant (Daffada and Dowsett 1995) was expressed in all of the aforementioned tissues except for liver.

The role of alternatively spliced variants in hormone-independent growth remains uncertain, in particular since it has been difficult to demonstrate stable variant receptor proteins. It is well known that there are substantial differences in tissue sensitivities to oestrogens and pharmacological agents, such as selective ER modulator (SERM). The mechanism underlying the differential agonist and antagonist activity in different tissues of SERM is not well understood. If the variants were expressed to the same extent as the wild-type ER, then they could potentially exert a differential influence on tissue sensitivity and on specificity of pharmacological and physiological agents. Recently, a "raloxifene response element" on the promoter region of at least one oestrogen-sensitive gene has been described (Yang et al. 1996). When the ER interacted with this element, via modulating proteins, it did not require the conventional DNA-binding region. Hence the Δ3 variant might interact with this pathway but not be able to bind to conventional EREs.

Furthermore, a number of coactivators and corepressors have been isolated recently which interact with ER (Horwitz et al. 1996). The binding of the coactivators enhances ER transcriptional activity, while the binding of corepressors reduces transcription. Variants which can dimerise with the wild-type ER might affect the binding of these coactivators or corepressors, or the homodimerisation of the variant receptor might compete with the wild-type ER for these coactivators or corepressors.

Thus whilst there is now a large amount of literature describing ER mRNA variants, there are few persuasive data indicating their biological importance. Data on protein expression has long been awaited and is required to clarify their importance.

11.6 Oestrogen Receptor β

All the work mentioned above relates to ERα. A second ER gene was recently cloned called ERβ. It contains all six domains of ERα and shares considerable homology in the DNA- and ligand-binding domain (Kuiper et al. 1996). The relative binding of the in vitro-translated protein has shown that ERβ can bind to oestradiol with the same affinity as ERα and can heterodimerise with ERα (Cowley et al. 1997). The role of ERβ in tamoxifen-resistant disease remains unknown.

11.7 Conclusion

Point mutations in ER can have profound effects on protein function in model systems. However, their relevance to breast cancer appears to be slight, in terms of both risk of development and phenotype of established tumours. A large number of variant ERs exist at the mRNA level. The lack of data on expression at the protein level leaves only circumstantial evidence of their clinical importance. There are a large number of data sets which are consistent with their involvement in the establishment of certain pathological tumour phenotypes, although not in the response of the tumour to tamoxifen. Theoretically, their differential expression could be important as an additional control on oestrogen/antioestrogen sensitivity. It does not appear likely that there will be any advantage to the measurement of any of these mutants or variants in routine clinical practise.

References

Allegra JC, Barlock A, Huff KK, Lippman ME (1980) Changes in multiple or sequential oestrogen receptor determinations in breast cancer. Cancer 45:792–794

Andersen TI, Wooster R, Laake K, Collins N. Warren W, Skrede M, Roz E, Tveit KM, Johnston SRD, Dowsett M, Olsen A, Moller P, Stratton MR, Borresen-Dale A-L (1997) Screening for ESR mutations in breast and ovarian cancer patients. Human Mutat 9:531–536

Castles CG, Fuqua SAW (1996) Alterations within the estrogen receptor in breast cancer. In: Pasqualini JR, Katzenellenbogen BS (eds) Hormone-dependent cancer. Dekker, New York, pp 81–106

Castles CG, Fuqua SAW, Klotz DM, Hill SM (1993) Expression of a constitutively active estrogen receptor variant in the estrogen receptor-negative BT-20 human breast cancer cell line. Cancer Res 53:5934–5939

Chan CMW, Dowsett M (1997) A novel oestrogen receptor variant mRNA lacking exons 4 to 6 in breast carcinomas. J Steroid Biochem Mol Biol 62:419–430

Cowley SM, Hoare S, Mosselman S, Parker MG (1997) Estrogen receptor α and β form heterodimers on DNA. J Biol Chem 272:19858–19862

Daffada AAI, Dowsett MD (1995) Tissue-dependent expression of a novel splice variant of the human oestrogen receptor. J Steroid Mol Biol 55:413–421

Daffada AAI, Johnston SRD, Smith IE, Detre S, King N, Dowsett M (1995) Exon 5-deletion variant oestrogen receptor mRNA expression in relation to tamoxifen resistance and PgR/pS2 status in human breast cancer. Cancer Res 55:288–293

DeFriend DJ, Howell A, Nicolson RI, Anderson E, Dowsett M, Mansel RE, Blamey RW, Bundred NJ et al (1994) Investigation of a new pure antiestrogen (ICI 182780) in women with primary breast cancer. Cancer Res 54:408–414

Dotzlaw H, Alkhalaf M, Murphy LC (1992) Characterization of estrogen receptor variant mRNAs from human breast cancers. Mol Endocrinol 6:773–785

Fuqua SAW, Wolf D (1995) Molecular aspects of estrogen receptor variants in breast cancer. Breast Cancer Res Treat 35:233–241

Fuqua SAW, Fitzgerald SD, Chamness GC, Tandon AK, McDonnell DP, Nawaz Z, O'Malley BW, McGuire WL (1991) Variant human breast tumor estrogen receptor with constitutive transcriptional activity. Cancer Res 51:105–110

Fuqua SAW, Fitzgerald SD, Allred DC, Elledge RM, Nawaz, McDonnell DP, O'Malley BW, Greene GL, McGuire WL (1992) Inhibition of estrogen receptor action by a naturally occurring variant in human breast tumors. Cancer Res 52:483–86

Fuqua SAW, Wiltschke C, Castles C, Wolf D, Allred DC (1995) A role for estrogen-receptor variants in endocrine resistance. Endoc Relat Cancer 2:19–25

Horwitz KB, Jackson TA, Bain DL, Richer JK, Takimoto GS, Tung L (1996) Nuclear receptor coactivators and corepressors. Mol Endocrinol 10:1167–1177

Hughes MR, O'Malley BW (1991) Genetic defects in receptors involved in disease. In: Parker MG (ed) Nuclear hormone receptors: molecular mechanisms, cellular functions and clinical abnormalities. Academic, London, pp 321–353

Johnston SRD, Saccaci-Jotti G, Smith IE, Newby J, Dowsett M (1995) Change in oestrogen receptor expression and function in tamoxifen-resistant breast cancer. Endocr Relat Cancer 2:105–110

Johnston SRD, Lu B, Dowsett M, Liang X, Kaufmann M, Scott GK, Osborne CK, Benz CC (1997) Comparison of oestrogen receptor DNA binding in untreated and antioestrogen-resistance human breast tumours. Cancer Res 57:3723–3727

Katzenellenbogen BS, Montano MM, Ekena K, Herman ME, McInernery EM (1997) Antiestrogens: mechanism of action and resistance in breast cancer. Breast Cancer Res Treat 44:23–38

Koehorst SGA, Jacobs HM, Thijssen JHH, Blankensteiln MA (1993) Wild type and alternatively spliced estrogen receptor messenger RNA in human meningioma tissue and MCF7 breast cancer cells. J Steroid Biochem Mol Biol 45:227–233

Korach KS (1994) Insights from the study of animals lacking functional estrogen receptor. Science 226:1524–1527

Kuiper GGJM, Enmark E, Pelto-Huikko M, Nilsson S, Gustafsson J-Å (1996) Cloning of a novel estrogen receptor expressed in rat prostate and ovary. Proc Natl Acad Sci USA 93:5925–5930

Kuukasjarvi T, Kononen J, Helin H, Holli K, Isola J (1996) Loss of estrogen receptor in recurrent breast cancer is associated with poor response to endocrine therapy. J Clin Oncol 14:2584–2589

Leygue ER, Watson PH, Murphy LC (1996) Estrogen receptor variants in normal human mammary tissue. J Natl Cancer Inst 88:284–290

Mahfoudi A, Roulet E, Dawois S, Parker MG, Wahli W (1995) Specific mutations in the estrogen receptor change the properties of antiestrogens to full agonists. Proc Natl Acad Sci USA 92:4206–4210

McCarty KS Jr, Szabo E, Flowers JL, Cox EB, Leight GS, Miller L, Kaurath J, Soper JT, Budwit DA, Greasman WT, Seigler HF, McCarty KS (1986) Use

of a monoclonal anti-estrogen receptor antibody in the immunohistochemical evaluation of human tumors. Cancer Res [Suppl] 46:4244(s)–4248(s)
McGuire WL, Chamness GC, Fuqua SAW (1991) Estrogen receptor variants in clinical breast cancer. Mol Endocrinol 5:1571–1577
Montano MM, Ekena K, Krueger KD, Keller AL, Katzenellenbogen BS (1996) Human estrogen receptor ligand activity inversion mutants: receptors that interpret antiestrogens and estrogens as antiestrogens and discriminate among different antiestrogens. Mol Endocrinol 10:230–242
Murphy LC, Wang M, Coutt A, Dotzlaw H (1996) Novel mutations in estrogen receptor messenger RNA in human breast carcinomas. J Clin Endocrinol Metab 81:1420–1427
Murphy LC, Dotzlaw H, Leygue E, Douglas D, Coutts A, Watson PH (1997) Estrogen receptor variants and mutations. J Steroid Biochem Mol Biol 62:363–372
Nomura Y, Tashiro H, Shinozuka K (1985) Changes in steroid hormone receptor content by chemotherapy and/or endocrine therapy in advanced breast cancer. Cancer 55:546–551
Pfeffer U, Fecarotta E, Vidali G (1995) Co-expression of multiple estrogen receptor variant messenger RNAs in normal and neoplastic breast tissues and in MCF-7 cells. Cancer Res 55:2158–2165
Rea D, Parker MG (1996) Effects of an exon 5 variant of the estrogen receptor in MCF-7 breast cancer cells. Cancer Res 56:1556–1563
Robertson JFR (1996) Oestrogen receptor: a stable phenotype in breast cancer. Br J Cancer 73:5–12
Roodi N, Bailey LR, Kao W-Y, Verrier CS, Yee CI, Dupont WD, Parl FF (1995) Estrogen receptor gene analysis in estrogen receptor-positive and receptor-negative primary breast cancer. J Natl Cancer Inst 87:446–451
Saez RA, Osborne CK (1989) Hormonal treatment of advanced breast cancer. In: Kennedy BJ (ed) Current clinical oncology. Liss, New York, pp 163–172
Smith EP, Boyd J, Frank GR, Takahashi H, Cohen RM, Specker B, Williams TL, Lubahn DB, Korach KS (1994) Estrogen resistance caused by a mutation in the estrogen-receptor gene in a man. N Engl J Med 331:1056–1061
Wang Y, Miksicek RJ (1991) Identification of a dominant negative form of the human estrogen receptor. Mol Endocrinol 5:1707–1715
Wang M, Dotzlaw H, Fuqua SAW, Murphy LC (1997) A point mutation in the human estrogen receptor gene is associated with the expression of an abnormal estrogen receptor mRNA containing 69 novel nucleotide insertion. Breast Cancer Res Treat 44:145–151
Witschke C, Lemieux P, Wolf DM, Castles CG et al (1996) Isolation of receptor variant from premalignant breast lesions. Breast Cancer Res Treat 37 [Suppl] abstract 29

Wolf DM, Jordan VC (1994) The estrogen receptor from tamoxifen stimulated MCF7 tumor variant contains a point mutation in the ligand binding domain. Breast Cancer Res Treat 31:129–138

Yang NN, Venugopalan M, Hardikar S, Glasebrook A (1996) Identification of an estrogen response element activated by metabolites of 17β-estradiol and raloxifene. Science 273:1222–1225

Zhang QX, Borg A, Wolf DM, Oesterreich S, Fuqua SAW (1997) An estrogen receptor mutant with strong hormone-independent activity from a metastatic breast cancer. Cancer Res 37:1244–1249

12 Structure and Expression of the Androgen Receptor in Prostate Cancer

J. Trapman

12.1 Introduction

During embryogenesis and puberty the male sex hormones, or androgens, testosterone, and 5α-dihydrotestosterone, play a key role in the development of the male phenotype in genotypical 46, XY individuals. In adult males, androgens control the function and structure of the male urogenital tissues, including the prostate. In the prostate, the enzyme 5α-reductase type 2 is responsible for conversion of testosterone to the more potent 5α-dihydrotestosterone (Thigpen et al. 1993). The action of both testosterone and 5α-dihydrotestosterone is mediated by the intracellular androgen receptor.

Initially, the growth of the majority of prostate cancers appears to be androgen-dependent. Consequently, if surgical intervention is no longer possible because of metastatic disease, orchiectomy and other forms of endocrine manipulation are generally accepted forms of treatment. Although many forms of hormonal therapy have been developed over the last few decades, prostate tumors almost inevitably escape normal hormonal regulation and continue to grow in an apparently hormone-refractory process.

As a general concept it is supposed that the initiation and progressive growth of prostate tumors is caused by a cascade of genetic alterations, and further modulated by epigenetic events. One of the specific questions of prostate cancer development and progression concerns the role of the androgen receptor. In this chapter recent findings on androgen receptor functioning, and androgen receptor structure and expression in prostate cancer are presented.

12.2 Structure and Function of the Androgen Receptor

In order to understand the mechanism of action of the androgens, and to be able to interfere effectively with this mechanism, a detailed knowledge of the structure and mechanism of action of the androgen receptor is a prerequisite.

The androgen receptor is a member of the subfamily of steroid hormone receptors of the family of nuclear receptors, also encompassing the glucocorticoid receptor, progesterone receptor, mineralocorticoid receptor, and estrogen receptors (Beato et al. 1995; Gronemeyer and Laudet 1995). Steroid hormone receptors are characterized by a central DNA-binding domain, that targets the receptor to specific DNA sequences (hormone response elements), a carboxy-terminal ligand-binding domain, and a large amino-terminal domain (Fig. 1). The DNA-binding and ligand-binding domains of the various steroid hormone receptors show structural homology. The amino-terminal domain varies considerably in length and in amino acid composition between the various receptors.

Importantly, the androgen receptor gene maps to the X chromosome, implicating that in males mutations in the androgen receptor become directly manifest. The open reading frame is separated over eight exons. Exon 1 encodes the large amino-terminal domain, the DNA-binding domain is encoded by exons 2 and 3, and the information for the

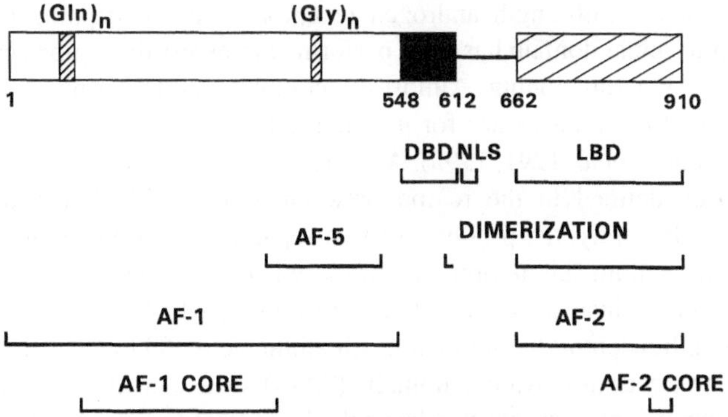

Fig. 1. The functional organization of the androgen receptor. The amino-terminal domain is shown as an *open box*, the DNA-binding domain (*DBD*) as a *black box*; the ligand-binding domain (*LBD*) as a *hatched box*. In the amino-terminal domain the Gln and Gly stretches are indicated. *NLS*, nuclear localization signal; *AF*, transactivation function

ligand-binding domain is distributed over exons 4–8 (Faber et al. 1989; Kuiper et al. 1989; Lubahn et al. 1989).

The amino-terminal domain of the androgen receptor contains two long homopolymeric stretches of amino acids, including long glutamine and glycine stretches, encoded by polymorphic CAG and GGC/T repeats, respectively (Faber et al. 1989). The CAG repeat, in particular, is extremely polymorphic, ranging from approximately 12–30 repeat units in healthy individuals. This offers the possibility of linkage studies in families with supposed defects in androgen receptor functioning. Because of the polymorphic repeats, the length of the androgen receptor can vary between 900 and 930 amino acid residues. This can lead to confusion concerning the numbering of the individual amino acid residues. In this chapter, the amino acid positions correspond to an androgen receptor composed of 910 amino acid residues [(Gln)20 and (Gly)16].

The amino-terminal domain of the androgen receptor, which has a length of over 500 amino acids, contains a strong transactivating function 1 (AF-1; see Fig. 1). It is supposed that this part of the receptor interacts with other specific and general transcription factors. These interactions might be direct, or can be mediated by co-activators. Starting from the full-length androgen receptor, deletion mapping of the amino-terminal domain has been performed to assess the regions essential for AF-1 functioning. Almost the complete amino-terminal region was found to be necessary for maximal AF-1 activity (Simental et al. 1991; Jenster et al. 1991, 1995). A core segment, with over 50% activity has been defined in the region between amino acids 101 and 307 (Fig. 1). Recently, the presence of two separate, but cooperating small segments within this region was proposed (Chamberlain et al. 1996). Interestingly, instead of the AF-1 core region, which is active in the full-length receptor, the segment from amino acid 360 to 485 functions as a strong transactivation domain (AF-5) in a constitutively active androgen receptor mutant that lacks the ligand-binding domain (Jenster et al. 1995). This observation indicated that the ligand-binding domain affects the properties of the amino-terminal domain, suggesting an interaction between the two domains. Recent functional in vivo protein–protein interaction studies with separate amino-terminal and ligand-binding domains demonstrated that such a ligand-dependent, functional interaction can indeed take place (Langley et al. 1995; Doesburg et al. 1997). However, it is unknown whether this interaction can also occur in the full-length receptor. Neither is it known whether this interaction is direct or indirect, or whether this proposed interaction in a full-length receptor is inter- or intramolecular.

A carboxy-terminal domain of approximately 250 amino acid residues contains the ligand-binding function of the androgen receptor. The integrity of the complete domain is important for steroid binding, because most deletions or point mutations in this region abolish proper ligand binding (Simental et al. 1991; Jenster et al. 1991; Quigley et al. 1995). A part of the ligand-binding domain is also important for receptor dimerization (Doesburg et al. 1997; Fawell et al. 1990). For most steroid hormone receptors, a transcription activation function (AF-2) in the ligand-binding domain has been demonstrated (Gronemeyer and Laudet 1995). This transactivation function depends on ligand binding for activity. Despite the high degree of structural homology in the core region of the AF-2 domain, no apparent AF-2 activity has so far been identified in the ligand-binding domain of the androgen receptor.

In the absence of ligand, the ligand-binding domain of the androgen receptor is associated with a large multiprotein complex, composed of heat shock proteins, that maintains the receptor in a transcriptionally inactive conformation. Heat shock proteins are supposed to be involved in proper folding, in intracellular trafficking, and in nuclear import of steroid hormone receptors (Smith and Toft 1993; Pratt 1993). Hormone binding leads to the dissociation of receptor-associated proteins and initiates conformational changes in the ligand-binding domain, which are important for subsequent receptor dimerization, DNA binding, and interaction with transcription mediators. In the absence of ligand, the androgen receptor is distributed over the nucleus and the cytoplasm. The addition of androgens rapidly induces a DNA-bound conformation (Jenster et al. 1993; Zhou et al. 1994). Nuclear translocation depends on an intact nuclear localization signal in the hinge region between the DNA-binding domain and ligand-binding domain.

Recently, the three-dimensional structures of the ligand-binding domains of two different retinoic acid receptors (RXR and RAR), the thyroid hormone receptor, and the estrogen receptor-α have been elucidated (Bourguet et al. 1995; Renaud et al. 1995; Wagner et al. 1995; Wurtz et al. 1996; Brozowski et al. 1997). They were shown to fold in 11 or 12 α-helices and two β-sheets, in a sandwich-like structure. The overall homology in the ligand-binding domains of the various nuclear receptors might be sufficient to predict the folding of the ligand-binding domain of the androgen receptor (Fig. 2). A common "mousetrap"-like mechanism of ligand binding has been proposed. By folding back to the core region of the ligand-binding domain, helix 12, which contains the AF-2 core sequence, comes into close contact with the ligand and seals the ligand pocket. The conformational change induced by hormone binding might create an interaction surface, which allows binding of transcriptional co-activators like the transcription intermediary factor and the glucocorticoid receptor-interacting protein (TIF-2/GRIP1), and the androgen receptor-specific co-activator Ara70 (Voegel et al. 1996; Hong et al. 1996; Yeh and Chang 1996).

Antagonists which inhibit the biological effects of androgens compete with agonists for binding to the receptor. As discussed in more detail in Sect. 12.5, antagonists are frequently applied in prostate cancer therapy. There is accumulating evidence that they induce a different conformational change upon binding to the ligand-binding domain (Zeng et al. 1994; Kuil et al. 1995). This aberrant conformation leads to a receptor with reduced, or without, transcriptional activity. Different antagonists may affect different aspects of receptor functioning, such as

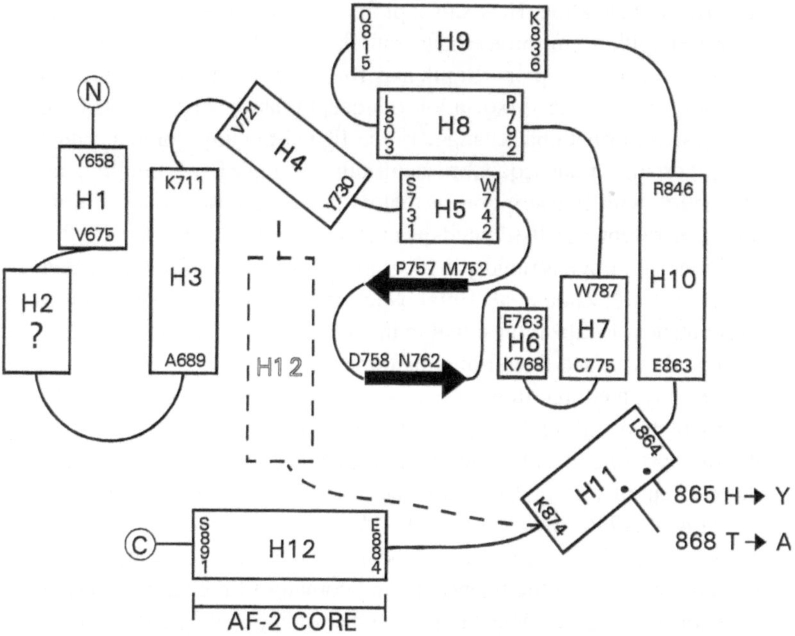

Fig. 2. The structure of the ligand-binding domain of the androgen receptor. α-Helices are indicated by *open boxes*. Positions of helices are deduced from the structure of other nuclear receptors. The proposed position of helix 12 in the presence of ligand, containing the transactivation function 2 (*AF-2*) core region, is within *broken lines*. *N*, amino terminus; *C*, carboxy terminus

dissociation of heat shock proteins, dimerization, DNA binding, interaction with general or specific transcription factors, interaction with co-activators, and interaction with the amino-terminal domain.

12.3 Androgen-Regulated Prostate Development

The important question regarding identification of androgen receptor target genes, which are essential for androgen-dependent growth of prostate cancer cells, has not been answered as yet. Many different genes have been found to be androgen-regulated in the human or rat prostate. It would not be surprising if androgen-regulated prostate cancer growth is a complex interplay between epithelial and stromal cell-driven processes. Part of these might be identical to mechanisms involved in normal prostate development.

The androgen receptor is a key factor in the development of the male reproductive tissues, including the prostate. During the initial steps of development, the androgen receptor is expressed in the mesenchymal cells of the embryonic anlagen of the reproductive tissues, in the case of the prostate the urogenital sinus (Cunha et al. 1987; Cooke et al. 1991; Bentvelsen et al. 1995). During early stages of development, androgen receptor expression is undetectable in the epithelial cells. However, it is well established that epithelial cell differentiation is driven by androgens. This implies that processes regulated by the activated androgen receptor in the mesenchymal cells direct epithelial cell differentiation. Later steps in prostate development might depend on the androgen receptor in the epithelial cells, eventually leading to the production of prostate-specific proteins (Donjacour et al. 1993).

The molecular mechanisms underlying androgen-regulated mesenchymal–epithelial cell interactions in the developing prostate are not well understood. It is assumed that androgen-regulated growth factors and extracellular matrix components produced by the mesenchymal cells act specifically on the epithelial cells, stimulating their proliferation and differentiation in a paracrine mechanism. One candidate androgen-regulated gene in prostate development is keratinocyte growth factor (KGF or FGF7), a member of the fibroblast growth factor family. KGF is expressed in the mesenchymal cells and its specific receptor, a splice variant of the FGF2 receptor, in the epithelial cells of the developing prostate. In rats, KGF expression is androgen-regulated and can at least partially replace androgens in in vitro prostate development (Yan et al. 1992; Sugimura et al. 1996). However, in KGF knock-out mice, prostate development appears to be normal, indicating redundancy in this pathway of prostate development (Guo et al. 1996).

12.4 Androgen Receptor Mutations in Androgen Insensitivity and Kennedy's Disease

Mutations in the androgen receptor gene have been found in three human diseases. The best documented mutations in the androgen receptor are those which are the cause of X-linked androgen insensitivity. Androgen insensitivity is a hereditary defect in male sexual development. The complete form of the androgen insensitivity syndrome is characterized by an apparently female phenotype, combined with a 46,XY karyotype and a normal or slightly elevated testosterone production (reviewed in Quigley et al. 1995). Partial impairment of androgen receptor functioning can result in an aberrant male phenotype or a predominantly female phenotype.

Over 100 different mutations in the androgen receptor have been described in androgen insensitivity (Gottlieb et al. 1997). The majority of these mutations are missense mutations, leading to an amino acid substitution in the DNA-binding or ligand-binding domain. As far as characterized, all of these mutations result in the inhibition of one or more functions of the receptor, including ligand binding and DNA binding. In the androgen insensitivity syndrome, mutations in the large exon 1, encoding the complete amino-terminal transactivation domain, are rare. The few mutations found in the amino-terminal domain are all nonsense mutations or frame shift mutations, leading to the synthesis of a truncated receptor. These findings indicate that the transactivating function of the amino-terminal domain is more flexible than the functions of other parts of the protein. Alternatively, in the case of mutation, the function can easily be taken over by another part of the protein, leading to a completely functional receptor.

Kennedy's disease, or SBMA (spinal and bulbar muscular atrophy) is characterized by a progressive degeneration of specific motor neuron cells. In SBMA, the length of the poly(Gln) stretch in the androgen receptor is expanded to more than 40 residues (LaSpada et al. 1991). Depending on the cell type and promoter tested, the expanded Gln stretch has been shown to cause a slight decrease or has no effect on the transactivation activity of the androgen receptor (Jenster et al. 1994; Kazemi-Esfarjani et al. 1995). Choong et al. (1996) recently reported that the expanded CAG repeat reduces androgen receptor mRNA and protein expression, and does not alter its functional activity. It is most likely, however, that expansion of the Gln stretch leads to a gain of a so far unidentified function in the target cells or to the deposition of specific complexes, which are the cause of the disease.

12.5 The Androgen Receptor in Prostate Cancer

Therapy of metastasized prostate cancer is in many cases based on the interference of androgen-regulated gene expression, by inhibition of testosterone production (orchiectomy, LHRH analogues), or by the blocking of androgen receptor functioning (anti-androgens). Although apparently effective for a short period, local recurrences and metastases develop over time. To be able to improve hormonal therapy, it is essential to elucidate the mechanisms of hormone-dependent and subsequent hormone-independent prostate tumor growth.

In general, immunohistochemical studies have demonstrated that the androgen receptor is expressed in the tumor cells; expression is variable in the stromal cell compartment (Van der Kwast et al. 1991; Sadi et al. 1991; Chodak et al. 1992; Sadi and Barrack 1993; Ruizeveld de Winter et al. 1994; Tilley et al. 1994; Pertschuk et al. 1995). Thus, androgen-regulated growth of the tumor might depend on both paracrine and autocrine androgen-directed mechanisms. So far, a role of stromal cell-driven, androgen-regulated processes in the proliferation of the epithelial tumor cells has not been studied in detail. In model systems, mesenchymal cells derived from the urogenital sinus, or stromal cells from the mature prostate, are able to support growth of human prostate tumor cells in male nude mice, whether or not the tumor cells express the androgen receptor (Chung et al. 1991). Evidence against a role of stromal cells in androgen-regulated prostate tumor growth is the observation that, in in vitro cell culture, proliferation of the LNCaP prostate tumor cell line is androgen-sensitive in the absence of stromal cells (Schuurman et al. 1988).

A second important issue concerns the role of the androgen receptor in apparently hormone-refractory prostate cancer. Obvious explanations for hormone-resistant tumor growth include bypassing of the androgen-receptor regulated pathway of prostate growth by an androgen-independent route, or an outgrowth of a tumor cell sub-population which was androgen-independent from the beginning. However, several points of evidence also implicate the androgen receptor as a component in at least part of the resistant tumors. Although more heterogeneous than in the normal prostate, the vast majority of resistant, locally progressive prostate and metastatic tumors show high levels of androgen receptor expression (Koivisto et al. 1997; Hobisch et al. 1995). The predominant nuclear localization of the receptor is in favor of an active conformation. In part of the local recurrencies, androgen receptor mRNA expression is higher than in the tumor prior to therapy (e.g., orchiectomy) (Koivisto et al. 1997). Interestingly, in bone metastases, which represent a late stage of hormone-refractory prostate cancer, the androgen receptor expression level seems even higher and more homogeneous than in the local, recurrent tumor (Kleinerman, unpublished).

A mechanism leading to androgen receptor over-expression can involve the amplification of the androgen receptor gene (Koivisto et al. 1997; Visakorpi et al. 1995). This mechanism has been observed in approximately 30% of recurrent hormone-refractory prostate tumors after orchiectomy. Importantly, androgen receptor gene amplification was not detected in tumors prior to endocrine therapy. Furthermore, amplification of the androgen receptor gene was predominantly found in tumors that relapse relatively late after the start of therapy. Combined with the survival data, these findings suggest that the tumors with an amplified receptor gene primarily respond to hormone therapy, followed by an escape mechanism including amplification and a reactivated androgen receptor. It remains to be established whether, in tumors without androgen receptor gene amplification, presumed androgen receptor over-expression results from other mechanisms of up-regulation of gene expression.

Another mechanism which could explain a functionally active androgen receptor in hormone therapy-resistant prostate cancer is the occurrence of a structural alteration in the receptor, as a result of an amino acid substitution due to a mutation in the gene. As the classical example, the androgen receptor in the LNCaP prostate cancer cell line contains a missense mutation in the ligand-binding domain, resulting in a Thr to Ala substitution at position 868. This amino acid substitution renders the receptor, and as a consequence the growth of LNCaP cells, responsive to anti-androgens (OH-flutamide and cyproterone acetate), and low-affinity ligands (estradiol, progesterone, and anandron) (Veldscholte et al. 1990). However, the situation concerning mutations in the androgen receptor in prostate tumor tissues is rather confusing. Different reports describe apparently conflicting data (Ruizeveld de Winter et al. 1994; Newmark et al. 1992; Culig et al. 1993; Suzuki et al. 1993; Gaddipati et al. 1994; Peterziel et al. 1995; Taplin et al. 1995; Tilley et al. 1996). Point mutations in the androgen receptor gene have been detected in primary and metastatic tumors with a varying frequency. However, the effect of many of these mutations on the transactivating function of the receptor has not been established. For a limited number of mutations in the ligand-binding domain, it has been shown that they correspond to a loss of ligand specificity, although in most cases the effects are less dramatic than for the Thr-Ala868 substitution. In these selected cases (Val706Met, Val721Met, His865Tyr, Thr868Ala/Ser) a mutation can lead to receptor activation by anti-androgens or by low-affinity ligands, like adrenal androgens (Veldscholte et al. 1990; Culig et al. 1993; Peterziel et al. 1995; Taplin et al. 1995; Tan et al. 1997; Fenton et al. 1997). Interestingly, amino acid substitutions His865Tyr and Thr868Ala/Ser are both in helix 11 (see Fig. 2). At present, the most likely explanation for androgen receptor mutations in late-stage prostate cancer is that they represent a mixture of functional mutations and random mutations resulting from genetic instability.

An alternative mechanism of ligand-independent activation of the androgen receptor has been demonstrated in a model system, DU145 cells, co-transfected with an androgen receptor expression plasmid and different androgen-regulated reporter genes (Culig et al. 1994). In this system, the androgen receptor can be activated by the growth factors insulin-like growth factor (IGF)-1 and KGF. KGF-activated androgen receptor seems promoter-dependent. Cross-talk with other signal transduction pathways has also been reported.

In conclusion, although indirect evidence is suggestive of an activated androgen receptor during endocrine therapy of prostate cancer, its mechanism of action and clinical relevance remain to be determined. Elucidation of the mechanisms of androgen-regulated development of the normal prostate and the growth of androgen-dependent prostate cancer can contribute to a better understanding of a presumed function of the androgen receptor in hormone-independent cancer, and to an improvement of prostate cancer therapy.

References

Beato M, Herrlich P, Schutz G (1995) Steroid hormone receptors: many actors in search of a plot. Cell 83:851–857

Bentvelsen FM, Brinkmann AO, Van der Schoot P, Van der Linden JETM, Van der Kwast ThH, Boersma WJ, Schroeder FH, Nijman JM (1995) Developmental pattern and regulation by androgens of androgen receptor expression in the urogenital tract of the rat. Mol Cell Endocrinol 113:245–253

Bourguet W, Ruff M, Chambon P, Gronemeyer H, Moras D (1995) Crystal structure of the ligand-binding domain of the nuclear receptor RXR-α. Nature 375:377–382

Brozowski AM, Pike ACW, Dauter Z, Hubbard RE, Bonn T, Engstrom O, Ohman L, Greene GL, Gustafsson J-A, Carlquist M (1997) Molecular basis of agonism and antagonism in the oestrogen receptor. Nature 389:753–758

Chamberlain NL, Whitacre DC, Miesveld RL (1996) Delineation of two distinct type 1 activation functions in the androgen receptor amino-terminal domain. J Biol Chem 271:26772–26778

Chodak GW, Kranc DM, Puy LA, Takeda H, Johnson K, Chang C (1992) Nuclear localization of androgen receptor in heterogeneous samples of normal, hyperplastic and neoplastic human prostate. J Urol 147:798–803

Choong CS, Kemppainen JA, Zhou Z, Wilson EM (1996) Reduced androgen receptor gene expression with first exon CAG repeat expansion. Mol Endocrinol 10:1527–1535

Chung LWK, Gleave ME, Hsieh JT, Hong S-J, Zhau HE (1991) Reciprocal mesenchymal–epithelial interaction affecting prostate tumor growth and hormonal responsiveness. Cancer Survey 11:91–121

Cooke PS, Young P, Cunha GR (1991) Androgen receptor expression in developing male reproductive organs. Endocrinol 128:2867–2873

Culig Z, Hobisch A, Cronauer MV, Cato ACB, Hittmair A, Radmayr C, Eberle J, Bartsch G, Klocker H (1993) Mutant androgen receptor detected in an advanced stage prostatic carcinoma is activated by adrenal androgens and progesterone. Mol Endocrinol 7:1541–1550

Culig Z, Hobisch A, Cronauer MV, Radmayr C, Trapman J, Hittmair A, Bartsch G, Klocker H (1994) Androgen receptor activation in prostate tumor cell lines by insulin-like growth factor-1, keratinocyte growth factor, and epidermal growth factor. Cancer Res 54:5474–5478

Cunha GR, Donjacour AA, Cooke PS, Mee S, Bigsby RM, Higgins SJ, Sugimura Y (1987) The endocrinology and developmental biology of the prostate. Endocrine Rev 8:338–362

Doesburg P, Kuil CW, Berrevoets CA, Steketee K, Faber PW, Mulder E, Brinkmann AO, Trapman J (1997) Functional in vivo interaction between the amino-terminal transactivation domain and the ligand binding domain of the androgen receptor. Biochemistry 36:1052–1054

Donjacour AA, Cunha GM (1993) Assessment of prostatic protein secretion in tissue recombinants made of urogenital sinus mesenchyme and urothelium from normal or androgen insensitive mice. Endocrinol 132:2342–2350

Faber PW, Kuiper GGJM, van Rooij HCJ, Van der Korput JAGM, Brinkmann AO, Trapman J (1989) The N-terminal domain of the androgen receptor is encoded by one large exon. Mol Cell Endocrinol 61:257–262

Fawell SE, Lees JA, White R, Parker MG (1990) Characterization and colocalization of steroid binding and dimerization activities in the mouse estrogen receptor. Cell 60:953–962

Fenton MA, Shuster TD, Fertig AM, Taplin M-E, Kolvenbag G, Bubley GJ, Balk SP (1997) Functional characterization of mutant androgen receptors from androgen-independent prostate cancer. Clin Cancer Res 3:1383–1388

Gaddipati JP, McLeod DG, Heidenberg HB, Sesterhenn IA, Finger MJ, Moul JW, Srivastava S (1994) Frequent detection of codon 877 mutation in the androgen receptor gene in advanced prostate cancers. Cancer Res 54:2861–2864

Gottlieb B, Trifiro M, Lumbroso R, Pinsky L (1997) The androgen receptor gene mutations database. Nucleic Acids Res 25:158–162

Gronemeyer H, Laudet V (1995) Transcription factors 3: nuclear receptors. Protein Profile 11:1173–1308

Guo L, Degenstein L, Fuchs E (1996) Keratinocyte growth factor is required for hair development but not for wound healing. Genes Dev 10:165–175

Hobisch A, Culig Z, Radmayer C, Bartsch G, Klocker H, Hittmair A (1995) Distant metastases from prostate carcinoma express androgen receptor protein. Cancer Res 55:3068–3072

Hong H, Kohli K, Trivedi A, Johnson DL, Stallcup MR (1996) GRIP1, a novel mouse protein that serves as a transcriptional coactivator in yeast for the hormone binding domains of steroid receptors. Proc Natl Acad Sci USA 93:4948–4952

Jenster G, Van der Korput HAGM, Van Vroonhoven C, Van der Kwast Th, Trapman J, Brinkmann AO (1991) Domains of the human androgen receptor involved in steroid binding, transcriptional activation and subcellular localization. Mol Endocrinol 5:1396–1404

Jenster G, Trapman J, Brinkmann AO (1993) Nuclear import of the human androgen receptor. Biochem J 293:761–768

Jenster G, de Ruiter PE, Van der Korput HAGM, Kuiper GGJM, Trapman J, Brinkmann AO (1994) Changes in the abundancy of androgen receptor isotypes: effects of ligand treatment, glutamine-stretch variation and mutation of putative phosphorylation sites. Biochemistry 33:14064–14072

Jenster G, Van der Korput HAGM, Trapman J, Brinkmann AO (1995) Identification of two transcription units in the N-terminal domain of the human androgen receptor. J Biol Chem 270:7341–7346

Kazemi-Esfarjani P, Trifiro MA, Pinsky L (1995) Evidence for a repressive function of the long polyglutamine tract in the human androgen receptor: possible pathogenetic relevance for the (CAG)n-expanded neuropathies. Hum Mol Genet 4:523–527

Koivisto P, Kononen J, Palmberg C, Tammela T, Hyytinen E, Isola J, Trapman J, Cleutjens K, Noordzij A, Visakorpi T, Kallioniemi O-P (1997) Androgen receptor gene amplification: a possible molecular mechanism for androgen deprivation therapy failure in prostate cancer. Cancer Res 57:314–318

Kuil CW, Berrevoets CA, Mulder E (1995) Ligand-induced conformational alterations of the androgen receptor analyzed by limited trypsinization. J Biol Chem 270:27569–27576

Kuiper GGJM, Faber PW, Van Rooij HCJ, Van der Korput HAGM, Ris-Stalpers C, Klaassen P, Trapman J, Brinkmann AO (1989) Structural organization of the human androgen receptor gene. J Mol Endocrinol 2:R1-R4

Langley E, Zhou ZX, Wilson EM (1995) Evidence for an anti-parallel orientation of the ligand-activated human androgen receptor dimer. J Biol Chem 270:29983–29990

LaSpada AR, Wilson EM, Lubahn DB, Harding AE, Fischback KH (1991) Androgen receptor mutations in X-linked spinal and bulbar muscular atrophy. Nature 352:77–79

Lubahn DB, Brown TR, Simental JA, Higgs HN, Migeon CJ, Wilson EM, French FS (1989) Sequence of the intron/exon junctions of the coding region of the human androgen receptor gene and identification of a point mutation in a family with complete androgen insensitivity. Proc Natl Acad Sci USA 86:9534–9538

Newmark JR, Hardy O, Tonb DC, Carter BS, Epstein JI, Isaacs WB, Brown TR, Barrack ER (1992) Androgen receptor gene mutations in human prostate cancer. Proc Natl Acad Sci USA 89:6319–6323

Pertschuk LP, Schaeffer H, Feldman JG, Macchia RJ, Kim Y-D, Eisenberg K, Braithwaite LV, Axiotis CA, Prins G, Greene GL (1995) Immunostaining for prostate cancer androgen receptor in paraffin identifies a subset of men with poor prognosis. Lab Invest 73:302–305

Peterziel H, Culig Z, Stober J, Hobisch A, Radmayr C, Bartsch G, Klocker H, Cato ACB (1995) Mutant androgen receptors in prostate cancer distinguish between amino acid sequence requirements for transactivation and ligand binding. Int J Cancer 63:544–550

Pratt WB (1993) The role of heat-shock proteins in regulating the function, folding, and trafficking of the glucocorticoid receptor. J Biol Chem 268:21455–21458

Quigley CA, De Bellis A, Marschke KB, El-Awady MK, Wilson EM, French FS (1995) Androgen receptor defects: historical, clinical and molecular perspectives. Endocrine Rev 16:271–321

Renaud JP, Rochel N, Ruff M, Vivat V, Chambon P, Gronemeyer H, Moras D (1995) Crystal structure of the RAR ligand binding domain bound to all-trans retinoic acid. Nature 378:681–689

Ruizeveld de Winter JA, Janssen PJA, Sleddens HMFB, Verleun-Mooijman MCT, Trapman J, Brinkmann AO, Santerse AB, Schroeder F, Van der Kwast ThH (1994) Androgen receptor status in localized and locally progressive hormone-refractory human prosate cancer. Am J Pathol 144:735–746

Sadi MV, Barrack ER (1993) Image analysis of androgen receptor immunostaining in metastatic prostate cancer. Cancer 71:2574–2580

Sadi MV, Walsh PC, Barrack ER (1991) Immunohistochemical study of androgen receptors in metastatic prostate cancer. Comparison of receptor content and response to hormonal therapy. Cancer 67:3057–3064

Schuurman AL, Bolt J, Voorhorst MM, Blankenstein RA, Mulder E (1988) Regulation of growth and epidermal growth factor receptor levels of LNCaP prostate tumor cells by different steroids. Int J Cancer 42:917–922

Simental JA, Sar M, Lane MV, French FS, Wilson EM (1991) Transcriptional activation and nuclear targeting signals of the human androgen receptor. J Biol Chem 266: 510–518

Smith DF, Toft DO (1993) Steroid receptors and their associated proteins. Mol Endocrinol 7:4–11

Sugimura Y, Foster BA, Hom YK, Lipschutz JH, Rubin JS, Finch PW, Aaronson SA, Hayashi N, Kawamura J, Cunha GR (1996) Keratinocyte growth factor can replace testosterone in the ductal branching morphogenesis of the rat ventral prostate. Int J Dev Biol 40:941–951

Suzuki H, Sato N, Watabe Y, Masai M, Seino S, Shimazaki S (1993) Androgen receptor gene mutations in human prostate cancer. J Steroid Biochem Mol Biol 46:759–765

Tan J, Sharief J, Hamil KG, Gregory CW, Zang D-Y, Sar M, Gumerlock PH, deVere White RW, Pretlow TG, Harris SE, Wilson EM, Mohler JL, French FS (1997) Dehydroepiandrosterone activates mutant androgen receptors expressed in the androgen-dependent human prostate cancer xenograft CWR22 and LNCaP cells. Mol Endocrinol 11:450–459

Taplin M-E, Bubley GJ, Shuster TD, Frantz ME, Spooner AE, Ogata GK, Keer HN, Balk SP (1995) Mutation of the androgen receptor gene in metastatic androgen independent prostate cancer. N Engl J Med 332:1393–1398

Thigpen AE, Silver RI, Guileyardo JM, Casey ML, McConnel JD, Russel DW (1993) Tissue distribution and ontogeny of steroid 5α-reductase isozyme expression. J Clin Invest 92:903–910

Tilley WD, Lim-Tio SS, Horsfall DJ, Aspinall JO, Marshall VR, Skinner JM (1994) Detection of discrete androgen receptor epitopes in prostate cancer by immunostaining: measurement by color video image analysis. Cancer Res 54:4096–4102

Tilley WD, Buchanan G, Hickey TE, Bentel JM (1996) Mutations in the androgen receptor gene are associated with progression of human prostate cancer to androgen independence. Clin Cancer Res 2:277–285

Van der Kwast ThH, Schalken J, Ruizeveld de Winter JA, van Vroonhoven CCJ, Mulder E, Trapman J (1991) Androgen receptors in endocrine therapy resistant human prostate cancer. Int J Cancer 48:189–193

Veldscholte J, Ris-Stalpers C, Kuiper GGJM, Jenster G, Berrevoets C, Claassen E, van Rooij HCJ, Trapman J, Brinkmann AO, Mulder E (1990) A mutation in the ligand binding domain of the androgen receptor of human LNCaP cells affects steroid binding characteristics and response to anti-androgens. Biochem Biophys Res Commun 173:534–540

Visakorpi T, Hyytinen E, Koivisto P, Tanner M, Keinanen R, Palmberg C, Tammela T, Isola J, Kallioniemi OP (1995) In vivo amplification of the androgen receptor gene and progression of human prostate cancer. Nature Genet 9:401–406

Voegel JJ, Heine MJS, Zechel C, Chambon P, Gronemeyer H (1996) TIF2, a 160 kDa transcriptional mediator for the ligand-dependent activation function AF-2 of nuclear receptors. EMBO J 15:3667–3675

Wagner RL, Apriletti JW, McGrath ME, West BL, Baxter JD, Fletterick R (1995) A structural role for hormone in the thyroid hormone receptor. Nature 378:690–697

Wurtz JM, Bourguet W, Renaud JP, Vivat V, Chambon P, Moras D, Gronemeyer H (1996) A canonical structure for the ligand-binding domain of nuclear receptors. Nature Struct Biol 3:87–94

Yan G, Fukabori Y, Nikolaropoulos S, Wang F, McKeehan WL (1992) Heparin binding keratinocyte growth factor is a candidate stromal to epithelial cell andromedin. Mol Endocrinol 6:2123–2128

Yeh S, Chang C (1996) Cloning and characterization of a specific coactivator, ARA70, for the androgen receptor in human prostate cells. Proc Natl Acad Sci USA 93:5517–5521

Zeng Z, Allan GF, Thaller C, Cooney AJ, Tsai SY, O'Malley BW, Tsai MJ (1994) Detection of potential ligands for nuclear receptors in cellular extracts. Endocrinol 135:248–252

Zhou Z, Sar M, Simental JA, Lane MV, Wilson EM (1994) A ligand-dependent bipartite nuclear targeting signal in the human androgen receptor. J Biol Chem 269:13115–13123

Subject Index

Ernst Schering Research Foundation Workshop

Editors: Günter Stock
Ursula-F. Habenicht

Vol. 1 *(1991)*: Bioscience⇆Society – Workshop Report
Editors: D. J. Roy, B. E. Wynne, R. W. Old

Vol. 2 (1991): Round Table Discussion on Bioscience⇆Society
Editor: J. J. Cherfas

Vol. 3 (1991): Excitatory Amino Acids and Second Messenger Systems
Editors: V. I. Teichberg, L. Turski

Vol. 4 (1992): Spermatogenesis – Fertilization – Contraception
Editors: E. Nieschlag, U.-F. Habenicht

Vol. 5 (1992): Sex Steroids and the Cardiovascular System
Editors: P. Ramwell, G. Rubanyi, E. Schillinger

Vol. 6 (1993): Transgenic Animals as Model Systems for Human Diseases
Editors: E. F. Wagner, F. Theuring

Vol. 7 (1993): Basic Mechanisms Controlling Term and Preterm Birth
Editors: K. Chwalisz, R. E. Garfield

Vol. 8 (1994): Health Care 2010
Editors: C. Bezold, K. Knabner

Vol. 9 (1994): Sex Steroids and Bone
Editors: R. Ziegler, J. Pfeilschifter, M. Bräutigam

Vol. 10 (1994): Nongenotoxic Carcinogenesis
Editors: A. Cockburn, L. Smith

Vol. 11 (1994): Cell Culture in Pharmaceutical Research
Editors: N. E. Fusenig, H. Graf

Vol. 12 (1994): Interactions Between Adjuvants, Agrochemical and Target Organisms
Editors: P. J. Holloway, R. T. Rees, D. Stock

Vol. 13 (1994): Assessment of the Use of Single Cytochrome P450 Enzymes in Drug Research
Editors: M. R. Waterman,
M. Hildebrand

Vol. 14 (1995): Apoptosis in Hormone-Dependent Cancers
Editors: M. Tenniswood, H. Michna

Vol. 15 (1995): Computer Aided Drug Design in Industrial Research
Editors: E. C. Herrmann, R. Franke

Vol. 16 (1995): Organ-Selective Actions of Steroid Hormones
Editors: D. T. Baird, G. Schütz, R. Krattenmacher

Vol. 17 (1996): Alzheimer's Disease
Editors: J.D. Turner, K. Beyreuther, F. Theuring

Vol. 18 (1997): The Endometrium as a Target for Contraception
Editors: H.M. Beier, M.J.K. Harper, K. Chwalisz

Vol. 19 (1997): EGF Receptor in Tumor Growth and Progression
Editors: R. B. Lichtner, R. N. Harkins

Vol. 20 (1997): Cellular Therapy
Editors: H. Wekerle, H. Graf, J.D. Turner

Vol. 21 (1997): Nitric Oxide, Cytochromes P 450, and Sexual Steroid Hormones
Editors: J.R. Lancaster, J.F. Parkinson

Vol. 22 (1997): Impact of Molecular Biology
and New Technical Developments in Diagnostic Imaging
Editors: W. Semmler, M. Schwaiger

Vol. 23 (1998): Excitatory Amino Acids
Editors: P.H. Seeburg, I. Bresink, L. Turski

Vol. 24 (1998): Molecular Basis of Sex Hormone Receptor Function
Editors: H. Gronemeyer, U. Fuhrmann, K. Parczyk

Supplement 1 (1994): Molecular and Cellular Endocrinology of the Testis
Editors: G. Verhoeven, U.-F. Habenicht

Supplement 2 (1997): Signal Transduction in Testicular Cells
Editors: V. Hansson, F. O. Levy, K. Taskén

Supplement 3 (1998): Testicular Function:
From Gene Expression to Genetic Manipulation
Editors: M. Stefanini et al.